Genetics in Ophthalmology

Developments in Ophthalmology

Vol. 37

Series Editor *W. Behrens-Baumann*, Magdeburg

KARGER

Basel · Freiburg · Paris · London · New York ·
Bangalore · Bangkok · Singapore · Tokyo · Sydney

Genetics in Ophthalmology

Volume Editors *B. Wissinger*, Tübingen
Susanne Kohl, Tübingen
U. Langenbeck, Frankfurt a.M.

20 figures, 6 in color, and 12 tables, 2003

KARGER

Basel · Freiburg · Paris · London · New York ·
Bangalore · Bangkok · Singapore · Tokyo · Sydney

Dr. B. Wissinger
Universitätsklinikum
Tübingen
Augenklinik
Molekulargenetisches Labor
Auf der Morgenstelle 15
D–72076 Tübingen

Dr. Susanne Kohl
Universitätsklinikum
Tübingen
Augenklinik
Molekulargenetisches Labor
Auf der Morgenstelle 15
D–72076 Tübingen

Prof. Dr. U. Langenbeck
Institut für Humangenetik
Universitäts-Klinikum
Theodor-Stern-Kai 7
D–60590 Frankfurt a.M

Continuation of 'Bibliotheca Ophthalmologica', 'Advances in Ophthalmology', and 'Modern Problems in Ophthalmology'

Founded 1926 as 'Abhandlungen aus der Augenheilkunde und ihren Grenzgebieten' by *C. Behr*, Hamburg and *J. Meller*, Wien

Former Editors: *A. Brückner*, Basel (1938–1959); *H.J.M. Wewe*, Utrecht (1938–1962); *H.M. Dekking*, Groningen (1954–1966); *E.B. Streiff*, Lausanne (1954–1979); *J. François*, Gand (1959–1979); *J. van Doaesschate*, Utrecht (1967–1971); *M.J. Roper-Hall*, Birmingham (1966–1980); *H. Sautter*, Hamburg (1966–1980); *W. Straub*, Marburg a.d. Lahn (1981–1993)

Library of Congress Cataloging-in-Publication Data

Genetics in ophthalmology / volume editors, B. Wissinger, S. Kohl, U. Langenbeck,
p. ; cm. – (Developments in ophthalmology, ISSN 0250–3751 ; v. 37)
Includes bibliographical references and indexes.
ISBN 3–8055–7578–5 (hbk. : alk. paper)
1. Eye–Diseases–Genetic aspects. I. Wissinger, B. II. Kohl, S. III. Langenbeck, U. IV. Series.
[DNLM: 1. Eye Diseases–genetics. 2. Gene Therapy. 3. Genetic Diseases, Inborn. 4. Genetic Predisposition to Disease. 5. Molecular Biology. WW 140 G3278 2003]
RE906.G395 2003
617.7′042–dc21

2003045797

Bibliographic Indices. This publication is listed in bibliographic services, including Current Contents® and Index Medicus.

Drug Dosage. The authors and the publisher have exerted every effort to ensure that drug selection and dosage set forth in this text are in accord with current recommendations and practice at the time of publication. However, in view of ongoing research, changes in government regulations, and the constant flow of information relating to drug therapy and drug reactions, the reader is urged to check the package insert for each drug for any change in indications and dosage and for added warnings and precautions. This is particularly important when the recommended agent is a new and/or infrequently employed drug.

www.karger.com
Printed in Switzerland on acid-free paper by Reinhardt Druck, Basel
ISSN 0250–3751
ISBN 3–8055–7578–5

Contents

Wissinger B, Kohl S, Langenbeck U (eds): Genetics in Ophthalmology.
Dev Ophthalmol. Basel, Karger, 2003, vol 37, pp 1–15

A Glimpse into Genomeland

U. Langenbeck

Institute of Human Genetics, University Hospital,
Frankfurt am Main, Germany

Abstract

This selective review of current genetic paradigms and procedures, presented in the context of surprising discoveries from the entire field of clinical and experimental genetics, may serve as a primer for the in-depth reviews of this volume. The rapid progress in clinical and molecular genetics requires continuing education and self-study of practising physicians to keep abreast of the developments that form the expanding field of genetic and molecular medicine.

Introduction

In Johann Wolfgang Goethe's drama 'Faust', the scholar Wagner communicates to Dr. Faustus his somewhat naive excitement about the advances of modern science with the words 'wie herrlich weit wir's doch gebracht' (how gloriously we have progressed). It is difficult though to remain as sceptical about progress as Dr. Faustus.

With an attitude midway between Wagner and Dr. Faustus, instead, we may feel like little Alice burning with curiosity about a rabbit with a waistcoat pocket and a watch to take out of it (Carroll, 1865 [1]). Such sort of rabbit was the famous problem that Dr. John Dalton (1766–1844), together with his eyes, had left to posterity, the causes of the anomaly which now carries his name. The day after his death, his vitreous humor was found to have no bluish hue, and looking through one of his eyes from behind did not yield false colours either. Both these findings refuted Dalton's predictions. Later speculations on the type of his colour blindness were finally solved by DNA analysis in parts of his eyes: Dalton was 'a deuteranope with a single long wave (LW) opsin gene that coded for a visual pigment with the shorter spectral sensitivity of the two common LW variants' [2]. This is an instructional example of how in the sciences curiosity may pay with spectacular findings.

Progress of Ophthalmogenetics, as Illustrated by the Blepharophimosis Trait

As an illustration of less spectacular yet steady progress I shall outline the history of blepharophimosis, because many paradigms of modern genetics and ophthalmogenetics are exemplified in this entity (see McKusick's catalogue of Mendelian genes and diseases [3, 4] with Nos. #110100, *601649 and *605597 and references cited therein).

The blepharophimosis trait is mainly a consequence of eyelid dysplasia. It consists of small palpebral fissures, epicanthus inversus, ptosis of eyelids and depressed nasal bridge, hence the acronym BPES (*b*lepharophimosis-*p*tosis-*e*picanthus inversus *s*yndrome). The earliest known reports are from 1875 (Galezowsky) and 1889 (Vignes), and the autosomal-dominant inheritance of BPES was clearly demonstrated in 1921 with a large pedigree from the USA. It became increasingly clear that the syndrome is associated in the majority of families with female sterility (primary or secondary amenorrhoea, premature ovarian failure (POF)), such that the trait is transmitted through males only. This pattern is denoted type I. Families with fertile female members are denoted type II.

The presence or absence of female sterility has been explained either by variable expressivity of a pleiotropic gene or by the action of different alleles. Furthermore, type I BPES was considered a 'contiguous gene syndrome' with a deletion of both the BPES gene and a linked POF gene. It has not been tested yet whether there are varying degrees of female subfertility in type II families as suggested by W. Lenz ('Lenz son') in 1985 [5]. Such a continuum eventually would cause false classifications in small families.

As in a great number of other genes, clinical cytogenetics has contributed to the chromosomal mapping of the BPES gene. Carriers of balanced translocations with breakpoints in 3q23, e.g. t(3;21)(q23;q22.1), and patients with deletions spanning 3q22.3–3q23.1 were reported to manifest the features of BPES. Linkage analysis in families with type I as well as with type II BPES again pointed to the 3q22–q23 region. There, through a series of pathogenic mutations, the causative gene, FOXL2, was finally identified. It is L subfamily member 2 of the forkhead/winged-helix family of *fo*rkhead homeobo*x* transcription factors [6], first described in *Drosophila*.

An exemplary genotype-phenotype analysis [7] has demonstrated that truncating FOXL2 mutations are associated with type I, whereas mutations

C-terminal to the forkhead domain cause BPES type II. Thus, both types are allelic. A second BPES locus, called 'BPES3', has been mapped through linkage analysis to the short arm of chromosome 7 (7p21–p13).

Type II BPES may be considered a non-syndromic blepharophimosis, i.e. a bodily localized trait, whereas type I BPES is a true syndrome with pleiotropic expression of the mutations. In addition to type I BPES, there is a greater number of blepharophimosis syndromes, documented e.g. in the London Dysmorphology Database (LDDB) [8]. Therein, the combination of the search terms 'blepharophimosis/blepharospasm', 'ptosis of eyelids', and 'epicantic folds' yields ten more established syndromes, and six unique patterns of malformation.

The presence of inactivating mutations as well as the 3q23 deletions indicate that haploinsufficiency is behind the mechanisms of BPES pathogenesis. W. Lenz [5] considered it possible that, as a corollary to the stiff face, a connective tissue anomaly of the theca ovarii might prevent ovulation. This charming hypothesis became sort of corroborated by the finding that the mouse FOXL2 gene is selectively expressed in the mesenchyme of the developing eyelids and in adult ovarian follicles.

Beyond this simple introductory example, the present statistics of ophthalmogenetic disorders appears quite daunting: Searching OMIM [4] for the term 'Eyes' yields 1,212 entries. And Winter's diagnostic LDDB [8] lists under the general term 'Eyes, globes' 1,755 syndromes and still unclassifiable single case reports. Already the sheer numbers testify to an impressive progress in ophthalmogenetics.

Of Genomes and Genes

Evolutionary Classification of Genes

Genetics is a universal science that embraces all of biology. In this vein, F. Lenz ('Lenz father') wrote in 1936 [9]: 'Rules that are obeyed the same way in peas and snapdragons, in flies and butterflies, in mice and rabbits, of course also apply to humans.' A few decades later, the knowledge of general mechanisms of gene action and regulation made possible the famous remark (ascribed to Jacques Monod): 'What is true for *E. coli* is true for elephants, only more so.'

This confidence into the unity of life also feeds the research on animal models of human diseases. A prominent example is the Pax6 gene: Human aniridia, the mouse mutant 'small eye' (*Sey*), and the 'eyeless' (ey) phenotype of *Drosophila* all concern this gene which controls eye development in all metazoans including the jellyfish, an animal with eyes but without a brain [10, 11]. These organisms share the same ancestry and the Pax6 genes are said to be related through speciation. Such genes are called orthologues. If, at some stage of evolution, genes are duplicated, they have a paralogous relationship to each other. When clusters of paralogues, e.g. the crystallin gene clusters, are found in a group of related organisms, it may become difficult to tell which cluster members are true orthologues [12].

Genomes are continuously rearranged during evolution through reciprocal translocations, inversions, and chromosome fusions and fissions. Fixation of such changes in an emerging species is a slow process, however. Sequenced pieces of the genome of one species may therefore be sorted along the finished genome sequence of a related species. The first such example is the nearly completed physical map of the mouse genome [13], where the draft sequence of the human genome [14, 15] served as template.

Building a physical genome map starts from a comprehensive sample of overlapping large-insert bacterial clones (cosmids and bacterial artificial chromosomes). These clones are arranged to form contigs. A minimal set of overlapping contigs is finally arranged into a 'tile path' and assigned to the various chromosomes. The final physical map of the mouse genome (representing the physis, not the physics of mice) is made up of 296 contigs and almost 17,000 sequence markers. The map is tied to the human draft sequence with 51,486 'homology crosslinks'. This high degree of homology allowed to fill a considerable number of gaps in the human sequence by data from the mouse map and vice versa.

Mouse and man still share 167 regions of conserved synteny (syntenic means 'together on the same strand'). These regions contain, in both species, the same set of orthologous genes, eventually with minor rearrangements within a region. Such rearrangements, mostly inversions, define a total of 288 conserved segments wherein the ordering of genes is identical in both species [13].

The synteny group around the Pax6 orthologues of mouse and man extends from human chromosome 11p14.1 to 11q12.2 [14] and in mouse chromosome 2 from locus 62.0 to 47.0 cM (centimorgan, distance from centromer in units of percent recombination) [16]. This area accommodates at least 180 genes, with some 50 of them identified, including the Wilms' tumour gene WT1 and the catalase gene CAT (mouse *Cas1*). A deletion of the region from Pax6 to WT1 is found in patients with the WAGR (*W*ilms' tumour, *a*niridia, *g*enitourinary abnormalities, mental *r*etardation) syndrome [4, No. #194070]. Another widely known region of conserved synteny is the distal end of mouse chromosome 16 which is homologous to most of the long arm of human chromosome 21 [17]. The conservation of synteny allows the prediction of gene locations and the homologization of diseases, from mouse to man and vice versa [18], and it has made possible the study of the pathogenesis of chromosome disorders. An example is mouse trisomy 16 as a model of Down syndrome [19].

Verification of Disease Genes and Pathogenic Mutations

A gene is identified as causing a monogenic disease if a pathogenic mutation co-segregates with the disease in a number of families. Hidden behind this seemingly simple definition are the complexities of finding a disease gene and of proving its functional defect.

Genetic diseases were studied in the past by either one of two ways, forward genetics or reverse genetics [20]. In the forward route, the disease was first delineated through pathological, histological, physiological, clinical-chemical and biochemical abnormalities. These abnormalities eventually localized the ultimate cause of the disease in a defined protein. Sequencing this protein, at least in part, allowed the cloning of the disease gene. This could take many years, gyrate atrophy with hyperornithinaemia due to ornithine transaminase deficiency [4, No. *258870] being an example.

The second way of delineating a hereditary disease, now called positional cloning, typically starts with a linkage analysis in as many afflicted families as possible, or with patients carrying a defined chromosome anomaly (deletions, translocations). These analyses define a chromosomal region (the disease locus) which still may contain dozens of different genes. With chromosome walking and jumping techniques, gene after gene is analysed for disease-specific organ and tissue expression and thereafter for possible pathogenic mutations. When the gene has been found, the pathophysiology and pathobiochemistry can be studied in a straightforward fashion, so as to 'explain' the disease. An example for this sequence of discoveries is choroideremia [4, No. #303100].

The modern combination of these two strategies is the candidate gene approach. It has been made possible through the success of the human genome project with an increasing number of annotated and functionally characterized genes. The pathophysiology of the disease or the gene content of the disease locus may immediately hint at a smaller number of genes as 'positional or functional candidates' for causing the disease [21]. This strategy has triggered the ever-accelerating pace of disease gene discovery.

The proof of a mutation as 'disease-causing' [22] is settled if a functional gene (not a pseudogene) is inactivated by an intragenic deletion, translocation or inversion or by a non-sense, frameshift or splice site mutation. Doubtful are the simple amino acid exchanges (missense mutations) because they could well represent functionally innocuous polymorphisms with allele frequencies between 0.01 and 0.99. Attempts to overcome this problem are segregation analysis (only the affected family members carry the mutation), by evolutionary comparison (the wild-type amino acid is conserved in a wide range of species, implying functional significance), and by demonstrating the absence of the mutant in at least 100 alleles of healthy controls (rendering the possibility of a common polymorphism unlikely). The ultimate proof comes from functional characterizations with the finding of a defective or aberrant expression or function of the mutant allele and its product, e.g. in transfected cell lines [23].

Of Proteomes and Proteins

Classical biological chemistry was and is concerned with the study of single proteins and their piecemeal re-assembly into enzyme complexes, biochemical pathways and signal transduction cascades. Molecular biology extended these studies to the machinery of protein synthesis, with analysis of gene sequences, of messenger RNA processing and of transcriptional and translational control. Molecular biologists also developed the tools for DNA sequencing and thus prepared the stage for the various genome projects and the field of genomics. The goal of genomics is to know the overall content, structure and evolution of genomes. So, with the completed draft sequence of the human genome 'in hand' [14, 15], we are now able to predict the number and in part also the function of our genes. Genomic techniques, e.g. array technologies, yield data on gene expression during normal development and in disease states. However, 'in the wonderland of complete sequences there is much that genomics cannot do, and so the future belongs to proteomics: the analysis of complete complements of proteins' [24]. If we are to know the true identities, functions, quantities, cellular localizations, post-translational modifications and mutual interactions of all the proteins in all cell types we have got to turn to proteomic techniques. Thus, proteomics has returned to biochemistry, now with factory-like approaches however, with large-scale, high-throughput and automated experiments, with *in silico* bioinformatic searches, with the full technical armament of genomics.

In its first phase, proteomics, like genomics [25], had been working in the cataloguing mode, for permitting researchers to pinpoint molecular abnormalities in disease states. Catalogues have been prepared, e.g. for isolated cell types, like photoreceptors [26], for histologically homogeneous structures, like the lens and cornea [27], or for isolated subcellular fractions [28].

The proteins (up to 2000) from such preparations are separated according to iso-electric point and molecular weight, respectively, by two-dimensional gel electrophoresis. The protein spots are isolated and then proteolytically digested within gel. The peptides of each protein are separated by liquid chromatography and identified by high-resolution mass spectrometric procedures with somewhat weird names, like MALDI-TOF MS (*m*atrix *a*ssisted *l*aser *d*esorption *i*onisation – *t*ime *of f*light *m*ass *s*pectrometry).

This inventory-like first-generation proteomic method allowed the analysis of candidate genes, e.g., in mouse cataracts [27]. Catalogues, however, cannot provide functional information. Tandem affinity purification is a second-generation, functional proteomics technique to overcome this problem by unravelling the networks of labile multiprotein complexes ('protein machines'). It combines genomic labelling with classical biochemical purification methods, mass spectrometry and bioinformatic analysis. This method has led in yeast to the definition of 232 distinct and validated multiprotein complexes [29]. And it is of prime medical interest because defects of these 'machines' may cause groups of related diseases, as suggested by different genetic types of Fanconi anaemia [4, No. *227650].

Modes of Mutant Gene Expression and Inheritance

Monogenic Inheritance

Formerly, in pre-genomic times, the rules of monogenic inheritance could be summarized as follows: Dominant inheritance is due to heterozygosity in a single gene and recessive inheritance is due to homozygosity, also in a single gene. If both parents are afflicted with the clinically identical recessive disease and nevertheless have healthy children they are homozygous in different genes. Alternatively, if all their children have the same disease they are homozygous for the same gene. Dominant and recessive diseases are due to mutations in different genes, and heterozygotes for autosomal recessive diseases are unaffected clinically. If a patient with a recessive disease has a child with the same disease, the spouse is a heterozygote and the pattern is called pseudo-dominance.

Finally, the 'one gene-one polypeptide' hypothesis tended to imply that mutations in a given gene will lead to only a single disease. The general validity of the latter assumption was already refuted in the 1960–70s with the various haemoglobin and glucose-6-phosphate dehydrogenase diseases.

All these simple rules and inferences may be found true in cases even today. They are studded, however, with exceptions, because they are built on the assumption of linear and additive interactions between the different genes and between the different cellular proteins.

Recessive diseases are caused by deficiency alleles (e.g. deletions, frameshift mutations, 'early' non-sense mutations), when, in the heterozygotes, the 50% residual amount or activity of protein (enzyme) is still sufficient to support the stability of a cellular or extracellular structure or the flux of metabolites through a biochemical pathway. A dominant disease will result when, with such mutations, the residual 50% no longer suffice to support these cellular processes. This cause of dominant diseases is called 'haplo-insufficiency'. If an aberrant protein disturbs the assembly of supramolecular structures we speak of 'dominant-negative' gene actions. If a mutation abolishes means of down-regulating gene expression, we have a 'toxic gain of function'. If a gene product is regulated very tightly it may become 'dose-sensitive'. In this case, too little as well as too much of protein will cause a dominant disease.

Quite obviously, the mode of inheritance is not a property of the mutation itself but a consequence of how the mutation is expressed in the context of existing structural and regulatory networks. Therefore, different mutations in a single gene may be transmitted either as a dominant trait, e.g. in the case of a dominant-negative mutation, or recessively when the mutation confers instability to the protein thus mimicking a deletion or a frameshift mutation.

Digenic Inheritance

If, after a local gene duplication event, the parent gene A and its linked duplicate B diverge to some degree in sequence but continue to act in the same histological or cellular structures or pathways, they may still cause the same disease, and AaBb double heterozygotes would be affected the same way as mutant homozygotes or compound heterozygotes of either a or b. An example is the DFNB1 locus for autosomal recessive deafness with its constituent connexin genes 26 (Cx26, GJB2) and 30 (Cx30, GJB6). The most prevalent mutations are a deletion of a single guanine (35del G) in Cx26 and the deletion of the whole Cx30 gene (del (GJB6-D13S1830)), respectively [30]. Because the Cx26 and Cx30 genes are tightly linked, they behave like a single gene and the recurrence risk for children of a DFNB1 patient and his unrelated spouse is negligible.

This recurrence risk will rise to 25%, however, when the two genes recombine freely, e.g. in digenic retinopathia pigmentosa (RP) with involvement of the homologous but unlinked peripherin/RDS and ROM1 genes [31]. In small families and without molecular investigations this pattern, for statistical reasons, may mimic pseudo-dominant inheritance (but with 25% instead of 50% recurrence risk). If the parents of a patient are unaffected, the disease risk for their further children is 25%, be the two contributing genes linked or not.

An extension of these recessive digenic biallelic cases is the recessive digenic triallelic model as described for some families with Bardet-Biedl syndrome (BBS) [32]. Mutant homozygotes in one BBS locus (BBS2) are affected only if they are, in addition, heterozygous at a second BBS locus (BBS6). The theoretical recurrence risk for sibs under this model is only 12.5%.

The pairs of contributing genes in these examples either code for homologous proteins that are part of the same multiprotein complexes (DFNB1, digenic RP) or they possibly act in the same developmental pathway (BBS). They are modifying each other's activity, and we speak of *digenic* inheritance because the contributing parental genotypes will not, in isolation, cause the disease. This concept is watered down, however, when 'digenic inheritance' is invoked in cases where a modifier gene alters the expression of an inherited disease [33].

Modifying effects of functionally and/or spatially related genes appear to be the rule rather than the exception, and most Mendelian diseases will end up as heterogeneous and complex traits [34, 35].

Mouse Models of Human Eye Diseases

Objectives and Obstacles

The last common ancestor of mouse and man lived some 70 million years ago, and dramatic changes in body size and brain function have occurred since. In contrast, little has changed with the biochemical pathways, if we disregard our inabilities to synthesize ascorbic acid or degrade uric acid. Conservation may also be assumed for the basic morphogenetic pathways, e.g. during development of hands and feet in man and paws in mice [36]. This relatedness in evolutionary terms of mice and men has made the mouse a favoured model for the study of the pathogenesis of human hereditary diseases and the first choice for evaluation of gene therapeutic protocols.

A note of caution, however, is appropriate: The same mutation (spontaneous and induced) may have different effects even in different inbred strains of mice [37, 38]. Therefore, an orthologous relationship of genes does not guarantee that identical mutations will recapitulate the human disease phenotype in all instances. Beside varying 'genetic backgrounds' of mice and men, gene recruitment in development and temporal and spatial gene regulation may have changed during evolution. 'A mouse is not a man, but it is as close as an experimental geneticist can get' [39].

Mouse Gene Nomenclature [40]

The (*italicized*) gene symbols start with capital letters for dominant traits and lower case for recessive ones. If a mutant trait turns out to be an allele of a main locus, it will be denoted as superscript of that locus. And if finally the gene has been identified, also the main locus' designation becomes a superscript together with the allele in question. This type of nomenclature is rather convenient for the historically minded because the classical allele names finally and visibly meet modern molecular knowledge.

Two examples may illustrate the rules. (1) The semidominant mutant Small eye (*Sey*), first described in 1967, maps to chromosome 2, as does the mutant Dickie's small eye (*Dey*) which was first described 11 years later. By crossing *Sey/+* and *Dey/+* heterozygous mice it was shown that *Dey* is an allele at the *Sey* locus, therefore the new symbol Sey^{Dey}.

Three independent Small eye alleles were shown in 1991 to result from mutations in the Pax-6 mouse homologue (orthologue), so the present symbol for 'Dickie's small eye' is *Pax6*$^{Sey\text{-}Dey}$. In the Neuherberg Institute of Mammalian Genetics (lab code: Neu) a Pax6 mutant with small eyes was obtained through ethyl nitrosourea (ENU) mutagenesis (see below) and got the designation *Pax6*$^{Sey\text{-}Neu}$. Nine more ENU-induced mutants at the *Pax6* locus received the superscripts *2Neu* to *10Neu*, with *Sey* omitted because the eye mutants were primarily linked to the *Pax6* locus. (2) When the dominant trait 'eye lens obsolescence' (*Elo*, do not confuse with the ELO trait of the human Diego blood group) turned out to be an allele at the second cataract locus, *Cat2*, it was denoted *Cat2*^{Elo}. We now know that the Cat2 locus harbours the structural gene for the crystallin γE chain, Cryge, so the present name of Elo is *Cryge*$^{Cat2\text{-}Elo}$. The Elo mouse carries in position 403 of the Cryge structural gene a single nucleotide (G) deletion, with the ensuing frameshift leading to a truncated γE crystallin protein [41]. Accordingly, an extended designation would be *Cryge, del 403 G (Cat2-Elo)*.

Spontaneous and Induced Mouse Mutants

Until 1936 [9], the mode of inheritance had been deduced empirically for a considerable number of human eye diseases, e.g. ocular albinism. Mice were already established experimental models for proving the Mendelian rules in mammals. Among the early themes in mouse genetics [37] were the 'dancing' phenotype and the diverse coat colours. Genetic eye diseases of the mouse first became known in 1924 (C. Keeler) with the 'rodless' retina mutation (r, now 'retinal degeneration', *Pde6b*rd), and in 1944 (H.B. Chase) with two non-allelic anophthalmic (eyeless) mutants (*Rax*^{ey1} and *ey2*). Today the number of defined loci and identified genes with phenotypic effects in the eyes (visual defects, morphological abnormalities and cornea and lens opacities) has grown to about 90 [16]. Nevertheless, with spontaneous mouse mutants only a small number of medically important genes may be detected. Therefore, mutagenesis screens with the potent mutagen ENU have been applied for increasing the number of mutant phenotypes in a larger number of genes [42]. Recently it has become possible to mass screen mice not only for absent or small eyes but also for more subtle eye diseases. These in vivo phenotype screens for eye and vision mutants include monitoring the head tracking response to a moving black-and-white grating, slit-lamp biomicroscopy for detection of anterior segment defects, and indirect ophthalmoscopy for the detection of retinal diseases [43]. We are witnesses of an emerging 'mouse ophthalmology'!

The mutual interest of mouse and human geneticists has synergistically enhanced the rapid progress of both disciplines. A report from a recent conference report [44] may serve as a final example: 'We saw pictures of hairless mice, fat mice, fat-free mice, cloned mice, mice with a Marfan syndrome analog, and many more. This proliferation of mice demonstrates the importance to human genetics of efficient research tools related to model organisms.'

And now, the smart zebrafish (*Danio rerio*) enters the field of ophthalmogenetics [45, 46]!

Databases for Clinical Use

'Online Mendelian Inheritance in Man (OMIM)', the Knowledge Base [4]

If you enter the office of a human geneticist and glance at his bookshelf there will be the chance to guess, from the number and different colours of the 'McKusick' volumes, how long he or she is in the field. The hard copy editions started white in 1966 with a mere 344 pages and ended with the two brown volumes of the 11th edition and a total of 4,335 pages in 1994 [3] when exponential growth of genetic and molecular knowledge had become unbearable for the print mode. Since then, because OMIM's delay of coverage is less than 2 weeks, it would be careless for a genetic counsellor not to consult OMIM (*http://www.ncbi.nlm.nih.gov/Omim)* before being consulted by his or her client.

Diagnostic Databases for PC

The LDDB [6] is structured by the symptoms' anatomical sites or tissues involved. There is a list of 14 diagnostic categories under the term 'Eyes, globes', and a list of seven under the term 'Eyes, associated structures'.

The diagnostic categories are subdivided into more detailed morphological terms, e.g., for the anterior chamber abnormalities we find 'anterior chamber abnormalities, unspecified' with 54, 'Peters' anomaly' with 18, and 'Rieger's anomaly' with 19 entries (syndromes and single case reports).

With the number of morphological terms given in parentheses, the following numbers of entries are listed for eye globes in the other diagnostic categories: conjunctiva (3): 49; cornea (8): 306; globes (8): 308; iris (7): 201; lens (3): 346; macula (2): 57; optic disc and nerve (4): 281; pupil (4): 66; retina (9): 369; sclera (2): 64; vision (9): 738; vitreous (2) 26.

Because a syndrome may affect more than a single domain of the eye globes, the sum of 2,865 well exceeds the number of 1,755 syndromes and still unclassifiable single case reports to which the general term 'Eyes, globes' is pointing. See *http://www.oup.com/uk/omd* for more details.

In addition to LDDB, many genetic and clinical centres also use the Australian (Agnes Bankier) 'Pictures of Standard Syndromes and Undiagnosed Malformations' (POSSUM) database which was launched in March 1987. Like

LDDB, this program uses of a hierarchical trait search list. The syndrome descriptions include the OMIM number, a list of synonyms, pictures of different patients at different ages, clinical and genetic comments, references, and a trait list. The program is continuously updated. The last version, 5.6, was released in May 2002. A platform-independent web version with links to OMIM will be launched for subscribers in October 2002. The pictures, however, will continue to be distributed on CD, see *http://www.possum.net.au* for further details.

Online Mutation Databases

Mutation databases serve a number of needs, e.g. for the study of mutation mechanisms and mutability or of population genetics and migratory history of man. For the clinical geneticist they facilitate comprehensive genotype-phenotype analyses and the study of variable clinical expression of given mutations. Mutation databases are either general (aiming at total coverage of genes) or locus-specific (aiming at total coverage of mutations).

The Human Gene Mutation Database at the Institute of Medical Genetics in Cardiff, UK, curated by D.N. Cooper and colleagues ([47], *http://www.hgmd.org/*), is the most useful general mutation database. Beside covering 1,163 genes with a total of 27,927 mutations (as of 21.06.2002), it presently has links to about 250 open locus-specific databases, e.g. for mutations in the retinoblastoma gene RB1 *(http:/www.d-lohmann.de/Rb/mutation.htm/)*.

The website of the HUGO Mutation Database Initiative [48], curated by RGH Cotton (Genomic Disorders Research Center, Melbourne, Australia) and the Human Genome Variation Society *(http:/www2.ebi.ac.uk/mutations/cotton)* contains a great number of links to locus-specific, central, general, national, and ethnic mutation databases. Also, articles from the Society Journal *'Human Mutation'* are freely available there, from vol. 7, 1996 on.

Outlook

As a schoolboy I enjoyed visiting the microscopic wonderland of salt crystals, starch grains and combs on ant legs. But the most marvellous object in my eyes was the water flea, *Daphnia*, with its heart beating 'in the wrong place', below its back. Only much later [49] I learned that such paradoxical findings in lobsters led Geoffroy de St. Hilaire in 1822 to his hypothesis of body plan inversion in vertebrates. Despite this (still unexplained) inversion, the establishment of the dorsal/ventral axis during ontogeny is conserved from the fruit fly *Drosophila* to the vertebrates, the orthologous genes having different names only [11].

Recourses to ontogeny and phylogeny have become cornerstones of genetic and molecular medicine as exemplified by the recent Nobel Prize in Physiology or Medicine honouring the worm *Caenorhabditis elegans* and its prophets Sydney Brenner, John Sulston and Robert Horvitz. Like the mouse (see above), the zebrafish, and the fly, the worm is serving the medical community with modules of evolutionary conserved ontogenetic mechanisms [46] that allow to establish developmental and functional models of human diseases. And it was the worm, which, through a model of ageing, taught us ways of how to get very old. In conclusion, it is heartening to see that, with the advent of genetic and molecular medicine, the medical specialities have acquired, for better communication, a common language and a successful new paradigm.

References

1 Carroll L (1865): Alice's Adventures in Wonderland. Reprint: Martin Gardner: The Annotated Alice. New York, Penguin Books, 1970.

2 Hunt DM, Dulai KS, Bowmaker JK, Mollon JD: The chemistry of John Dalton's color blindness. Science 1995;267:984–988.

3 McKusick VA: Mendelian Inheritance in Man. A Catalog of Human Genes and Genetic Disorders, ed 11. Baltimore, The Johns Hopkins University Press, 1994.

4 Hamosh A, Scott AF, Amberger J, Bocchini C, Valle D, McKusick VA: Online Mendelian Inheritance in Man (OMIM), a knowledgebase of human genes and genetic disorders. Nucleic Acids Res 2002;30:52–55, see *http://www.ncbi.nlm.nih.gov/Omim*

5 Lenz W: Zukunftsperspektiven in der Humangenetik; in Hammerstein W, Lisch W (eds): Ophthalmologische Genetik. Suttgart, Enke, 1985, pp 384–390.

6 Kaestner KH, Knöchel W, Martinez DE: Unified nomenclature for the winged helix/forkhead transcription factors. Genes Dev 2000;14:142–146.

7 De Baere E, Dixon MJ, Small KW, Jabs EW, Leroy BP, Devriendt K et al: Spectrum of FOXL2 mutations in blepharophimosis-ptosis-epicanthus inversus (BPES) families demonstrates a genotype-phenotype correlation. Hum Mol Genet 2001;10:1591–1600.

8 Winter R, Baraitser M: London Dysmorphology Database, vers 3.0. Oxford, Oxford University Press Electronic Publishing, 2001.

9 Lenz F: Die Methoden menschlicher Erbforschung; in Baur E, Fischer E, Lenz F (eds): Menschliche Erblehre, ed 4. München, 1936, pp 587–657.

10 Gehring WJ: The genetic control of eye development and its implications for the evolution of the various eye types. Int J Dev Biol 2002;46:65–73.

11 Wilkins AS: The Evolution of Developmental Pathways. Sunderland/Mass, Sinauer Assoc Inc, 2002.

12 Holland PW: The effect of gene duplication on homology. Novartis Found Symp 1999;222:226–242.

13 Gregory SG, Sekhon M, Schein J, Zhao S, Osoegawa K, Scott CE et al: A physical map of the mouse genome. Nature 2002;418:743–750.

14 International Human Genome Sequencing Consortium (Feb 15, 2001): Initial sequencing and analysis of the human genome. Nature 2001;409:860–921.

15 Venter JC, Adams MD, Myers EW, Li PW, Mural RJ, Sutton GG et al (Feb 16, 2001): The sequence of the human genome. Science 2001;291:1304–1351.

16 Mouse Genome Informatics, release 2.8, *http://www.informatics.jax.org/mgihome/* (June 28, 2002).

17 Mural RJ, Adams MD, Myers EW, Smith HO, Miklos GL, Wides R et al: A comparison of whole-genome shotgun-derived mouse chromosome 16 and the human genome. Science 2002;296:1661–1671.
18 Ehrlich J, Sankoff D, Nadeau JH: Synteny conservation and chromosome rearrangements during mammalian evolution. Genetics 1997;147:289–296.
19 Reeves RH, Baxter LL, Richtsmeier JT: Too much of a good thing: Mechanisms of gene action in Down syndrome. Trends Genet 2001;17:83–88.
20 McKusick VA: Mechanisms in genetic diseases of man. Am J Med 1957;22:676–686.
21 Blanton SH, Liang CY, Cai MW, Pandya A, Du LL, Landa B et al: A novel locus for autosomal-dominant non-syndromic deafness (DFNA41) maps to chromosome 12q24-qter. J Med Genet 2002;39:567–570.
22 Cotton RGH, Scriver CR: Proof of 'disease-causing' mutation. Hum Mutat 1998;12:1–3.
23 Scriver CR, Waters PJ, Sarkissian C, Ryan S, Prevost L, Cote D et al: PAHdb: A locus-specific knowledgebase. Hum Mutat 2000;15:99–104 *(http://data.mch.mcgill.ca/pahdb_new)*.
24 Fields S: Proteomics in genomeland. Science 2001;291:1221–1224.
25 Sohocki MM, Malone KA, Sullivan LS, Daiger SP: Localization of retina/pineal-expressed sequences: Identification of novel candidate genes for inherited retinal disorders. Genomics 1999;58:29–33.
26 Nishizawa Y, Komori N, Usukura J, Jackson KW, Tobin SL, Matsumoto H: Initiating ocular proteomics for cataloging bovine retinal proteins: Microanalytical techniques permit the identification of proteins derived from a novel photoreceptor preparation. Exp Eye Res 1999;69:195–212 (for an extended account, see Matsumoto H, Komori N: Methods Enzymol 2000;316:492–511).
27 Steely HT Jr, Clark AF: The use of proteomics in ophthalmic research. Pharmacogenomics 2000;1:267–280.
28 Morel V, Poschet R, Traverso V, Deretic D: Towards the proteome of the rhodopsin-bearing post-Golgi compartment of photoreceptor cells. Electrophoresis 2000;21:3460–3469.
29 Gavin AC, Bösche M, Krause R, Grandi P, Marzioch M, Bauer A et al: Functional organization of the yeast proteome by systematic analysis of protein complexes. Nature 2002;415:141–147.
30 Del Castillo I, Villamar M, Moreno-Pelayo MA, del Castillo FJ, Alvarez A, Telleria D et al: A deletion involving the connexin 30 gene in non-syndromic hearing impairment. N Engl J Med 2002;346:243–249.
31 Kajiwara K, Berson EL, Dryja TP: Digenic retinitis pigmentosa due to mutations at the unlinked peripherin/RDS and ROM1 loci. Science 1994;264:1604–1608.
32 Katsanis N, Ansley SJ, Badano Jl, Eichers ER, Lewis RA, Hoskins BE et al: Triallelic inheritance in Bardet-Biedl syndrome, a Mendelian recessive disorder. Science 2001;293:2213–2214.
33 Vincent AL, Billingsley G, Buys Y, Levin AV, Priston M, Trope G et al: Digenic inheritance of early-onset glaucoma: CYP1B1, a potential modifier gene. Am J Hum Genet 2002;70:448–460.
34 Scriver CR: Why mutation analysis does not always predict clinical consequences: Explanations in the era of genomics. J Pediatr 2002;140:502–506.
35 Badano JL, Katsanis N: Beyond Mendel: An evolving view of human genetic disease transmission. Nat Rev Genet 2002;3:779–789.
36 Gilbert SF: Developmental Biology, ed 6. Sunderland/Mass, Sinauer Assoc Inc, 2000.
37 Lyon MF, Rastan S, Brown SDM: Genetic Variants and Strains of the Laboratory Mouse, ed 3. Oxford, Oxford University Press, 1996, vol 1 & 2.
38 Gerlai R: Gene-targeting studies of mammalian behaviour: Is it the mutation or the background genotype? Trends Neurosci 1996;19:177–181.
39 Anderson KV: Finding the genes that direct mammalian development. ENU mutagenesis in the mouse. Trends Genet 2000;16:99–102.
40 International Committee on Standardized Genetic Nomenclature for Mice, Jackson I, Chairperson (2000): Rules and guidelines for gene, allele, and mutation nomenclature. *http://www.informatics.jax.org/mgihome/nomen/gene.shtml* (June 28, 2002).
41 Cartier M, Breitman ML, Tsui LC: A frameshift mutation in the γE-crystallin gene of the Elo mouse. Nat Genet 1992;2:42–45.
42 Rathkolb B, Fuchs E, Kolb HJ, Renner-Müller I, Krebs O, Balling R et al: Large-scale N-ethyl-N-nitrosourea mutagenesis of mice – From phenotypes to genes. Exp Physiol 2000;85:635–644.

43 Thaung C, West K, Clark BJ, McKie L, Morgan JE, Arnold K et al.: Novel ENU-induced eye mutations in the mouse: Models for human eye diseases. Hum Mol Genet 2002;11:755–767.
44 Lazzeroni LC, Karlovic CA: Genotype to phenotype: Associations, errors and complexity. Trends Genet 2002;18:283–284.
45 Vihtelic TS, Hyde DR: Zebrafish mutagenesis yields eye morphological mutants with retinal and lens defects. Vision Res 2002;42:535–540.
46 Shin JT, Fishman MC: From zebrafish to human: Modular medical models. Annu Rev Genomics Hum Genet 2002;3:311–340.
47 Krawczak M, Ball EV, Fenton I, Stenson PD, Abeysinghe S, Thomas N, Cooper DN: Human Gene Mutation Database – A biomedical information and research resource. Hum Mutat 2000;15:45–51.
48 Cotton RGH, McKusick V, Scriver CR: The HUGO Mutation Database Initiative. Science 1998;279:10–11.
49 Hogan BL: Upside-down ideas vindicated. Nature 1995;376:210–211.

Prof. Dr. Ulrich Langenbeck
Institute of Human Genetics, University Hospital
Theodor-Stern-Kai 7, D–60590 Frankfurt am Main (Germany)
Tel. +49 69 6301 6008, Fax +49 69 6301 6002, E-Mail u.langenbeck@em.uni-frankfurt.de

Wissinger B, Kohl S, Langenbeck U (eds): Genetics in Ophthalmology.
Dev Ophthalmol. Basel, Karger, 2003, vol 37, pp 16–33

Epidemiology of Hereditary Ocular Disorders

Thomas Rosenberg

Gordon Norrie Centre for Genetic Eye Diseases,
National Eye Clinic for the Visually Impaired, Hellerup, Denmark

Abstract

Molecular genetic evidence has contributed significantly to the understanding of the fundamental molecular biology of the eye in health and disease, but it has also exposed the inadequacy of our traditional clinically based classification of hereditary eye disorders by unravelling significant genetic non-allelic heterogeneity in many eye disorders. Furthermore, our understanding of the epidemiology of hereditary ocular disorders has gained considerably by the establishment of mutation spectra in a rapidly growing number of monogenic eye disorders. In this overview, special emphasis has been put on the growing impact of genetic eye diseases in visual impairment, genetic heterogeneity, and the role of founder mutations for the skewed appearance of certain hereditary ocular disorders in some populations.

Introduction

The human eye is a highly differentiated structure involving the concerted action of a large fraction of the total genetic make-up of a human being. The number of genes involved is echoed by a very large number of genetic disorders involving the eye, either as solely ophthalmic traits or as components of multi-system disorders.

Demographically, the prevalences and the diagnostic spectrum of hereditary eye diseases vary throughout the world, depending among other variables on population characteristics like size, age (generation number), growth, bottlenecks, migration, genetic isolation, and socio-cultural factors such as religion, political affairs, language, mating pattern, and social class.

In countries with high-level economies, a low prevalence of preventable and curable eye disorders is reflected in a proportionally high occurrence of genetically determined eye diseases.

As the genetic factor is obvious in most single-gene ocular disorders, this subgroup is often considered synonymous with the hereditary ocular disorders. Nevertheless, many of the numerically weighty ocular diseases, e.g. glaucoma, age-related macular degeneration, strabismus, and high myopia, exhibit complex aetiologies including genetic factors.

Population-based epidemiological figures are rare, due to the relative low prevalences of hereditary ocular disorders which presupposes large study populations. Thus, epidemiological data on hereditary ophthalmic disorders are often procured through studies from institutional sources, which generally are encumbered by incompleteness and selection bias. Furthermore, in some disorders, e.g. congenital cataract, microphthalmos, and developmental anomalies of the anterior chamber, the frequent occurrence of phenocopies hampers the aetiological classification of non-familial cases.

Another impediment is due to the ophthalmic diagnostic vocabulary itself, which to large extent has passed unchanged since the pioneering period of the speciality in the second half of the 19th century, implicating the use of symptomatic diagnoses, such as retinitis pigmentosa (RP), congenital nystagmus, and congenital cataract.

The low diagnostic specificity is mirrored by an extensive genetic heterogeneity in these and many other diagnostic groups. Ophthalmic literature is richly supplied with significant epidemiological contributions concerning hereditary disorders. A thorough account is beyond the scope of this overview, for which reason the readers are referred to ophthalmic genetic textbooks.

Recently, molecular genetic evidence has provided a new set of coordinates for the taxonomy of ocular disorders, which in the future has to be integrated with the existing diagnostic system, representing a great challenge to our differential diagnostic capabilities. So far, molecular genetic studies in a number of diseases, for example the Usher syndromes, the Bardet-Biedl syndromes, and Leber's congenital amaurosis, have demonstrated the existence of an almost perplexing genetic complexity leading to a wave of interest in refinement of differential diagnostic criteria.

The establishment of mutation spectra in different ocular disorders has not only clarified genotype-phenotype correlations, but has also been able to account for some epidemiological characteristics. Usually, autosomal dominant and X-linked disorders such as aniridia and ocular albinism, which show a relatively high proportion of new mutations, occur with similar frequencies in different populations. In general, these mutations arise in only one or a few families, and accordingly these disorders are characterized by large allelic heterogeneity. However, recurrent identical mutations do occur at so-called 'hot spots' in some disorders.

This presentation focuses on a few characteristic features of the epidemiology of hereditary ocular conditions exemplified by a few illustrative cases. In addition, some examples from the relatively well-characterized Danish population are mentioned.

The use of acronyms for genes and loci is in accordance with the OMIM nomenclature *(http://www.ncbi.nlm.nih.gov/Omim/)* as indicated in the text by italics.

The Role of Genetic Eye Diseases in Visual Impairment

The overall impact of heredity on the prevalence and appearance of ocular pathology remains obscure. A simple addition of the prevalences of the 20 most common single-gene ophthalmic disorders yield a proportion between 100 and 200 affected per 100,000 individuals. Figures on the relative frequency are available from statistics on visual impairment in various child populations [1]. The same statistics reveal interesting details with respect to the diversity in the nosological composition of the materials, which mainly are generated from blind registries or from special schools for the blind and partially sighted.

In the Nordic countries, genetic aetiology accounts for 31–40% [2, 3]. The corresponding prevalences of genetic eye disorders among all visually impaired children in four Nordic countries were between 19 and 44 per 100,000 children. The proportion of genetically determined conditions among children with non-systemic ocular disorders was significantly higher, i.e. 61%, despite a low consanguinity rate in these populations. The large difference is due to the fact that visual handicap associated with other disabilities secondary to brain disorders constitutes a large proportion of the registered cases in these populations. A study on the incidence of visual impairments among Nordic children in 1993 revealed 32% with a genetic aetiology in a study group, in which only 30% had non-systemic visual impairment [4].

In a large-scale investigation of childhood blindness caused by hereditary disease, Gilbert et al. [5] performed blind school studies from 13 countries in Latin America, Asia and Africa. In these countries, genetic disease was a major cause of childhood blindness affecting 11–39% of the pupils. In the eastern Mediterranean region, genetic disease was responsible for 59–80%. The highest prevalences were ascertained in countries with higher levels of socio-economic development and peaked in countries with high consanguinity rates. Accordingly, autosomal recessive conditions were common, counting retinal dystrophies such as RP and Leber's amaurosis as the most common diagnostic group. Regional variations were found, however, including high percentages of cataract and dislocated lenses in Malawi, Thailand and the Philippines, and albinism and microphthalmos as common conditions in Kenya and Uganda. In the economically most underdeveloped territories of Palestinian Arabs, the West Bank and Gaza Strip, hereditary disease accounted for 55% of childhood blindness, with Leber's amaurosis, degenerative myopia, and optic atrophy as the leading causes. The levels of consanguinity were 44 and 85%, respectively, in the two communities [6]. Among children with congenital cataract from 13 states of India, a major part of the cases was considered to be autosomal recessive due to a high incidence of parental consanguinity [7]. In contrast, the hereditary mode of non-systemic congenital cataract in Western Europe is autosomal dominant in the majority of genetically documented cases [8, 9].

Cross-sectional aetiological investigations among visually impaired populations including all age groups are scarce. In general, the relative frequency of single-gene disorders is weakened in the elderly due to the heavy impact from age-related diseases, such as macular degeneration and glaucoma. In a study from Newfoundland and Labrador, Green et al. [10] found that single-gene disorders accounted for 30% of the cases of total blindness. In the blind population of Belgium, probable and established genetic aetiology constituted the most important factor (43%) in adults aged 18–60 years. The corresponding fraction among older people was 23% [11].

In conclusion, single-gene ocular disorders are reported across the world as either the leading cause or among the most important causes of blindness and impaired vision in children and younger adults.

Genetic Heterogeneity

Genetic Classification According to the Mode of Inheritance

Prior to the molecular genetic era, genetic classification relating to the Mendelian modes of inheritance was of fundamental importance for genetic counselling, and still is, both with and without knowledge of the specific mutation involved.

RP is one of the most common genetic eye disorders with astonishing uniform prevalence rates between 20 and 40 per 100,000 in different populations [12–16]. A world-standardized prevalence rate of 22.4 per 100,000 was calculated in the Danish population [16]. Despite this conformity, the breaking-up into categories according to Mendelian mode of inheritance exposed considerable differences in genetic composition, especially for autosomal dominant and X-linked cases and families [13, 17–20]. Haim [20] suggested a reconsideration of the classification criteria, in order to create an exclusive and exhaustive definition fulfilling the demand of one, and only one, group for each family or patient. She performed an interesting reclassification of some published materials, and demonstrated that the differences observed were, to a certain extent, caused by different classification criteria. Among the problems associated with the categorization of families were insufficient family information, variable expressivity, and only a few generations available [21]. This problem, of course, is common to most hereditary human disorders with more than a single mode of inheritance, and is accentuated by an increasing amount of disorders with mitochondrial, digenic, and other non-Mendelian segregation patterns.

Non-Allelic Genetic Heterogeneity

Non-allelic genetic heterogeneity is rather the rule than the exception in ophthalmology. Despite sophisticated differential diagnostic tools for the phenotypic characterization of disorders with differential effects on the ocular morphology and the neurophysiology of vision, our clinical equipment is not yet able to reflect the molecular events behind many ocular disorders. As one of the most intensely studied disease groups, RP may again serve as a paradigm for non-allelic genetic heterogeneity (see also Hims et al., this issue).

The first RP gene to be discovered was the *RHO* gene encoding the rod visual pigment, rhodopsin [22]. This initiated a systematic and successful search for mutations in other genes encoding proteins involved in rod phototransduction and the rod visual cycle. It has now become evident that mutations in nearly all of these genes are responsible for larger or smaller fractions of the overall occurrence of RP [23]. Additional fractions of the RP group are due to mutations in cyto-skeletal genes [24, 25], in genes involved in intracellular trafficking [26–28], in transcription factors [29–31], or in genes implicated in the splicing machinery [32, 33]. So far, mutations in cloned genes are accounting for less than 50% of the total genetic composition of RP, even in the most thoroughly investigated populations. In cone or cone-rod dystrophies which are closely related to RP, 3 genes out of 10 known loci have been cloned, and in Leber's congenital amaurosis, which often is considered a subgroup of RP, 6 loci are known, out of which 4 represent cloned genes. The number of genetic RP loci will almost certainly rise in the future, and the final number of 'RP genes' will stay uncertain for the next few years, but may possibly amount to about 50. The genes involved in all kinds of hereditary retinal disorders are currently being monitored [34], and at present this website includes 134 genetic loci, including 89 cloned genes (February 21, 2003).

One of the possible explanations for the failure to demonstrate genetic heterogeneity clinically may be related to the fact that the coupling mechanism between the specific protein dysfunction and universal retinal cell destruction is still largely unknown. Provided the cell destruction is a common end result of various genetic mechanisms, clinical methods designed for the monitoring of abnormal visual functions may be largely irrelevant for the diagnosis of genetic heterogeneity.

Allelic Genetic Heterogeneity

Mutation screening in large panels of many ocular disorders has revealed a great variation regarding mutation type as well as the position of mutations. In general, nucleotide substitutions, deletions, as well as insertions are found in various proportions according to the different disorders. More complex molecular rearrangements, including more than one type of mutation, are also reported.

Some mutations result in functional null alleles, while other mutations induce more or less serious effects on gene expression and protein function. This diversity in mutation spectrum is reflected in the clinical picture, and many genotype-phenotype studies have established more or less firm correlations. More than 100 mutations in the rhodopsin gene have been reported, the majority of which are responsible for autosomal dominant RP. Other rhodopsin mutations are the cause of autosomal recessive RP [35], while a few specific mutations are associated with congenital stationary night blindness (CSNB) [36, 37]. The same has been shown to be the case for the β-subunit of the phosphodiesterase (*PDE6B* gene) [38], in which the majority of the mutations hitherto reported are found in families with autosomal recessive RP. A single exception is a His258Asn missense mutation responsible for autosomal dominant CSNB in a large multi-generation family [39].

The severity of the clinical phenotype is often partly determined by the location and type of the mutation. A striking example of a mild phenotype with onset in the late 20s was ascertained in a family with X-linked RP2, which normally is characterized by an onset in infancy and a relatively rapid visual decline [40].

In some disorders, as e.g. choroideremia, only functional null alleles are reported, leaving the possible effect of less severe mutations unknown [41]. In other disorders, exemplified by the Aicardi syndrome, crucial mutations in hemizygous males are unknown, which has been interpreted as incompatibility with life [42].

In the carbohydrate-deficient glycoprotein (CDG) syndrome type Ia due to mutations in the *PMM2* gene, encoding phosphomannomutase 2, the heterozygotic incidence of two mutations, 422G→A and 357C→A accounted for the majority of the examined cases in Northern Scandinavia [43]. In Denmark, the same base substitutions accounted for 86% of all mutations. Surprisingly, no patient was homozygous for either of the two common mutations, suggesting that homozygosity for these mutations is lethal [44]. In a Dutch-Danish study of CDG type Ia, an expected prevalence of the CDG syndrome of 1:20,000 was calculated based on allele frequencies in the two populations, pointing towards a significant number of undiagnosed cases with this condition [45].

The positions of disease-causing mutations vary in a non-random manner, including either exonic or intronic positions or more distant locations affecting upstream or downstream control loci. Many mutations are reported to cluster in important functional domains regulating gene-gene interaction, or to result in crucial protein alterations affecting for instance ligand-binding ability, post-translational modification, or degradation.

Among the most obvious clinical effects of allelic genetic heterogeneity is the occurrence of shared genetic backgrounds in clinically different ophthalmic disorders. Mutations in the ATP-binding cassette gene, *ABCA4*, are reported responsible for both Stargardt disease, fundus flavimaculatus, RP, and cone dystrophy [46], and certain mutations have also been related to the occurrence of age-related macular degeneration (see also Klaver and Allikmets, this issue) [47]. Another example is the *RDS* gene in which mutations are implicated in autosomal dominant RP and vitelliform macular dystrophy of late onset [48].

Mutations in the collagen matrix gene *TGFBI* have been found in different corneal dystrophies regarded as distinct clinical entities [49].

In *PAX6*, the so-called master gene of the eye, mutations are reported responsible for different anterior segment disorders, among others keratitis, atypical coloboma, Peters' anomaly, and aniridia [50]. *PAX6* mutations are also involved in the WAGR syndrome, a contiguous gene syndrome involving, among others, the *WT1* gene [51, 52]. Furthermore, the genotype of aniridia is correlated with the degree of iris hypoplasia; missense mutations are chiefly associated with the slightest iris anomalies, while total aniridia most often is associated with mutations blocking the function of the homeobox or other functional important domains [53].

Mutations affecting the fibrillin gene *FBN1* account for 40–50% of individuals fulfilling the diagnostic criteria of Marfan syndrome, a relatively frequent multi-system connective tissue disorder with ectopia lentis and high myopia affecting about 1 in 10,000, and characterized by a large inter- and intra-familial variation in phenotypic expression. So far, genotype-phenotype correlations have been slow to emerge, except for mutations in CaEGF motifs of exons 24–27 and 31–32, which are responsible for severe and neonatal phenotypes [54, 55].

Non-Mendelian Inheritance

Since the pioneering work by Wallace et al. [56], accumulating evidence have demonstrated the impact of mitochondrial DNA (mtDNA) mutations as the cause of neurometabolic disease, with Leber's hereditary optic neuropathy (LHON) as far the most common disorder (see also Howell, this issue). Three so-called 'primary' mtDNA point mutations, 11778G→A, 3460G→A, and 14484T→C, have been identified [57, 58]. The existence of a significant number of unaffected mutation carriers in the female lineages of LHON families indicates that a primary mtDNA mutation may be considered a necessary, but in many cases not a sufficient, prerequisite for disease expression. The factors responsible for this variation could be either genetic (nuclear or mitochondrial) or epigenetic [59].

LHON families without a confirmed primary mutation are also reported indicating further genetic heterogeneity [60].

The mitochondrial genome is exclusively maternally inherited and demonstrates no recombination events, creating a condition for phylogenetic population analyses. Haplotype analyses based on characteristic polymorphic restriction sites in mitochondrial genes, and polymorphisms in the non-coding D loop, have shown that the primary LHON mutations are of recent origin and have occurred recurrently on different haplotype backgrounds, while so-called 'secondary' LHON mutations are ancient polymorphisms with highly significant haplotype associations [61]. It has thus been argued that the expression of the primary LHON mutations 11778 and 14484 is associated with the J haplotype (Torroni nomenclature), which has been reported characteristic for the Caucasoid European population [59, 61, 62]. The German study group was also able to demonstrate a specific haplotype association of another mitochondrial disorder, Wolfram syndrome or DIDMOAD with optic atrophy as a characteristic component [62].

Among mitochondrial disorders with ophthalmic implications, chronic progressive external ophthalmoplegia and RP are most commonly associated with large-scale mtDNA deletions.

Chinnery et al. [63] published a retrospective population-based 10-year study from an integrated clinical and laboratory mitochondrial service in the North East of England, reporting on the collective epidemiology of pathogenic mtDNA mutations. They found a point prevalence of 6.6 per 100,000 individuals below pensionable age with confirmed disease caused by mtDNA defects. Furthermore, 7.6 per 100,000 individuals were at risk of developing mtDNA disease. Based on geographic differences within the area of investigation, the authors suggested that the degree of underdetection of manifest mitochondrial disease might be as high as one third [63]. Among the ascertained cases, 50% were due to LHON with a prevalence rate of 3.3 per 100,000.

Non-Mendelian inheritance has also been observed in digenic RP, involving the combined presence of pathogenetic *ROM1* and *RDS* mutations [64, 65]. Recently, an interesting paper appeared, demonstrating triallelic inheritance through the combined effect of two pathogenic *BBS2* alleles and a mutated

MKKS allele as a prerequisite for disease expression in some families with Bardet-Biedl syndrome (see also Koenig, this issue) [66]. This observation deserves special attention, due to its possible corollary for the genetic elucidation of complex traits.

Founder Effects in Ophthalmology

As a consequence of the selection of ophthalmic conditions which are either rare or of significant occurrence in a study population due to genetic drift, the literature on hereditary eye disorders may leave a false impression of the general frequency of the disorders in question. Among the European countries, the Finnish population is the most thoroughly studied with examples of more than a dozen so-called 'Finnish disorders', many of which are ophthalmic [67, 68].

The present population of Finland evolved from relatively few settlers of Baltic-Finnish and Germanic origin who immigrated into the southern and western part of the country about 2000 years ago. The central and northern part of the country was colonized less than 20 generations ago. Furthermore, until recently these settlers lived in many small communities, isolated by topography, large distances, and low population density [69].

Disorders with Autosomal Dominant Inheritance

Unusual high prevalences of rare autosomal dominant ophthalmic disorders in limited geographical regions may be due to single ancestors in large multi-generation families. CSNB in the Nougaret family of Southern France is a famous example of this, originally published by Cunier as early as in 1838 [70], and reinvestigated in 1907 [71]. Remarkable examples from the Nordic countries are Finnish amyloidosis [72], helicoid peripapillary chorioretinal dystrophy in Iceland [73], and Best macular dystrophy in Sweden [74, 75]. In these and other cases, subsequent linkage analyses initiated the final identification of the involved gene and the specific mutation.

Similar examples are known from Denmark, in which a large family with autosomal dominant CSNB was described in 1909, and rediscovered after 80 years [76], followed by the establishment of a new adCSNB locus, and the identification of a disease-associated mutation in the *PDE6B* gene [77].

Two extended pedigrees with congenital cataract, Marner cataract and Volkman cataract, led to the identification of two loci on chromosome 16 and chromosome 1, respectively [78, 79].

In genealogically unrelated families of geographically widespread occurrence, the presence of specific haplotype-mutation associations has been established in a number of autosomal dominant diseases as well.

In Sorsby's fundus dystrophy, Wijesuriya et al. [80] demonstrated a highly significant disease-associated haplotype spanning across 3 cM of the *TIMP3* locus in 68% of all affected chromosomes in patients from various parts of the British Isles, Ireland and the USA. Another British example of a founder effect is the recently cloned *NCL* gene, in which an S50T mutation is responsible for autosomal dominant RP [81].

In North Carolina macular dystrophy, a shared disease-associated haplotype was found in families from across the USA in North Carolina, South Carolina, Illinois, Wisconsin, West Virginia, and Texas, suggesting that most of the American families share the same ancestral mutation [82, 83].

In families with autosomal dominant optic atrophy in the UK, Votruba et al. [84] found a common disease-associated haplotype. This interpretation, however, has been questioned [85]. Thiselton et al. [86] rendered a common founder probable in 42% of the Danish families with the same disorder. The examples, mentioned above, underscore the significance of ancestral mutations, and furthermore illustrate that the genetic composition of many ophthalmic disorders with autosomal dominant transmission may exhibit considerable variation from place to place and from country to country. The majority of the known founder mutations seem to be of recent origin, i.e. 2–400 years or 10–20 generations ago. The geographical dispersal often bears witness to historical events such as colonization of new territories, migrations secondary to wars and persecution, urbanization, or simple adventuring.

X-Linked Disorders

X-linked conditions also offer striking examples of founder mutations. In incomplete CSNB, haplotype studies in descendants of the Mennonite population who immigrated to Alberta, Canada, were helpful in the cloning of the disease gene *CACNA1F* [87].

A significant high Finnish prevalence of choroideremia was elucidated prior to the identification of the *CHM* gene by segregation analysis and historical-genealogical evidence, demonstrating the presence of three large kindreds in remote Northern Finland probably originating from a common founder couple born in 1644 and 1646, 12 generations back [88]. Subsequently, *CHM* deletions were identified in these kindreds [89].

In X-linked retinoschisis (RS), a disorder with a strikingly high prevalence in three geographically separated clusters in Finland, Huopaniemi et al. [90], based on haplotype analyses, identified three mutations accounting for 95% of all Finnish RS cases. Each of these mutations was confined to one of the clusters. Nevertheless, the most common mutation did also occur in another study comprising 14% of all RS mutations reported from many European countries [91]. Based on genealogical studies, the involved mutations in Finland had estimated ages of >600, >300, and about 200 years. In contrast, the mutations in four additional Finnish families occurred in 1–5 generations, only [90].

In Denmark, a unique deletion was found in several RS families, accounting for nearly 50% of the registered RS cases in the country. Further investigations demonstrated a shared disease-associated haplotype, indicating an ancestral founder of the disease in these families [92].

Disorders with Autosomal Recessive Inheritance

Linkage disequilibrium mapping in founder populations has proved instrumental for the localization and identification of several disease genes in ocular disorders with autosomal recessive inheritance. Although founder mutations are frequently related to genetic isolates [68], particular mutations in some autosomal recessive ocular disorders are predominating and responsible for a large proportion of the diagnosed cases even in large panmictic populations.

The already mentioned, Finnish diseases [67] are examples of the former category, including among others cornea plana, Usher syndrome type 3, gyrate chorioretinal atrophy, neuronal ceroid lipofuscinosis infantile type, and eye-muscle-brain disease.

Usher syndrome offers striking examples of founder mutations in other genetic isolates as well. Among the Samaritans, an ancient community in Israel and Jordan, distinct with respect to religion and culture and a high frequency of consanguineous marriages, Usher syndrome type 1B is due to homozygosity for a single mutation in the *GARP* gene. A unique haplotype, found only in all USH1B carriers and affected individuals, implied that the disease-causing mutation, which was subsequently identified, probably came from a single founder [93, 94].

In the Acadian community of Southern Louisiana, Usher syndrome type 1 is more frequent than in any other place in the USA [95]. The Acadians emigrated from Normandy in France in the 17th century and settled in Quebec from where a few families migrated to Louisiana 7–8 generations ago. A mutation, 216G→A of the Acadian Usher type *USH1C* gene, was identified and was in complete linkage disequilibrium with an unique intronic expansion with 9 tandem repeats in Acadian patients only, implicating a common founder [96].

Pingelap Island of the Eastern Caroline Islands in the Pacific has become known as the 'island of the color blind' [97]. In this island a typhoon towards the end of the 18th century reduced the population to about a dozen individuals. Among the descendants of these survivors, achromatopsia occurs in 4–10%, the highest prevalence of a genetic disease ever reported. Homozygosity mapping identified an achromatopsia locus on chromosome 8q21–q22, and was followed by the identification of the *CNGB3* gene (see also Deeb and Kohl, this issue) [98]. Another famous achromatopsia island is the small island of Fur in

Northern Jutland, Denmark, which was visited by, among others, the European ophthalmic geneticists Franceschetti and Klein during the 1st International Congress of Human Genetics in Copenhagen, 1956 [99]. Further genealogical investigations have traced the ancestral lines of the Fur island achromats to two couples who lived in the second half of the 17th century. Recent molecular genetic analyses showed the homozygous presence of a common ancient founder mutation, 1148delC, in the *CNGB3* gene in a few residual patients from the original Fur island investigation [Sundin O, Wissinger B, pers. commun.].

Mutation analyses from larger and genetically mixed populations have revealed an intriguing homogeneity in a number of autosomal recessive ocular disorders with uniform prevalences. This is the case in the juvenile form of Batten disease in which a 1-kb deletion of the *CLN3* gene is responsible for nearly 90% of the cases in Europe.

Another striking example is the oculocutaneous albinism (OCA), which in all its variants has significant impact on visual function. In OCA1A a highly predominant *TYR* mutation, G47D, was found among Moroccan Jews. The same mutation on an identical haplotype background was found in patients from the Canary Islands and Puerto Rico, suggesting that the G47D mutation in these ethnically distinct populations may have a common origin [100].

OCA2 is a common genetic disorder on the African continent. Exceptionally high prevalences were found in Soweto, South Africa (1:3,900), Zimbabwe (1:4,728), Cameroon, Zaïre, and the Central African Republic. Haplotype analysis suggests that a highly prevalent 2.7-kb *OCA2* gene mutation arose before the divergence of these African populations which is estimated to about 2,000–3,000 years ago [101]. The high prevalence of OCA2 in these populations may indicate some selective factor or genetic drift, which has been further reinforced in the Tonga ethnic group, a genetic isolate in Zambesi Valley, Northern Zimbabwe, in which the prevalence of OCA2 was 1:1,000. [102].

In Usher syndrome type 2A with an estimated minimal prevalence of 1:30,000, the mutation 2299delG in the *USH2A* gene is prevailing in the European population. A single core-haplotype was found to be associated with this mutation, indicating an ancestral mutation that has spread throughout Europe and into the New World as a result of migration [103].

Concluding Remarks

At this time (2003), we are passing one of the milestones along the evolutionary road of scientific biological science. The identification of the genomic organization of living organisms has given us considerable insight into the epidemiology of disease-causing mutations and the structural genetic mechanisms behind disease expression. Much work is, indeed, ahead before we have gained a comprehensive picture of all disease genes and their prevalence in our populations. Today it may seem impossible to achieve such a goal in most countries of the world. Nevertheless, with the growing hope of future prevention and treatment of genetic disorders, specific knowledge of the underlying genetic cause might be mandatory for the introduction of these measures, and future technical improvements permitting large-scale mutation screening of DNA samples will eventually appear.

The post-genomic or proteomic era is just setting off, and may be expected to add new insight into the epidemiology of hereditary ocular disorders, a discipline which also in the future will form part of our common ophthalmic consciousness.

Acknowledgement

Søren Nørby, MD, PhD was of great help in valuable discussions and a critical reading of the manuscript.

References

1 Foster A, Gilbert C: Epidemiology of visual impairment in children; in Taylor D (ed): Paediatric Ophthalmology. Oxford, Blackwell Science, 1997, pp 3–12.

2 Rosenberg T, Flage T, Hansen E, Rudanko SL, Viggosson G, Riise R: Visual impairment in Nordic children. II. Aetiological factors. Acta Ophthalmol (Copenh) 1992;70:155–164.

3 Blohmé J, Tornqvist K: Visual impairment in Swedish children. II. Etiological factors. Acta Ophthalmol Scand 1997;75:199–205.

4 Rosenberg T, Flage T, Hansen E, Riise R, Rudanko SL, Viggosson G, Tornqvist K: Incidence of registered visual impairment in the Nordic child population. Br J Ophthalmol 1996;80:49–53.

5 Gilbert C, Rahi J, Eckstein M, Foster A: Hereditary disease as a cause of childhood blindness: Regional variation. Results of blind school studies undertaken in countries of Latin America, Asia and Africa. Ophthalmic Genet 1995;16:1–10.

6 Elder MJ, De Cock R: Childhood blindness in the West Bank and Gaza Strip: Prevalence, aetiology and hereditary factors. Eye 1993;7:580–583.

7 Vanita JR, Singh D: Genetic and segregation analysis of congenital cataract in the Indian population. Clin Genet 1999;56:389–393.

8 Francis PJ, Berry V, Bhattacharya SS, Moore AT: The genetics of childhood cataract. J Med Genet 2000;37:481–488.

9 McCarty CA, Taylor HR: The genetics of cataract. Invest Ophthalmol Vis Sci 2001;42:1677–1678.

10 Green JS, Bear JC, Johnson GJ: The burden of genetically determined eye disease. Br J Ophthalmol 1986;70:696–699.

11 Francois J, Haustrate-Gosset F: Causes of blindness in the Belgian adult population; in Eriksson AW, Forsius HR, Nevanlinna HR, Workman PL, Norio RK (eds): Population Structure and Genetic Disorders. London, Academic Press, 1980, pp 445–452.

12 Boughman JA, Conneally PM, Nance WE: Population genetic studies of retinitis pigmentosa. Am J Hum Genet 1980;32:223–235.

13 Boughman JA, Caldwell RJ: Genetic and clinical characterization of a survey population with retinitis pigmentosa. Prog Clin Biol Res 1982;82:147–166.

14 Bunker CH, Berson EL, Bromley WC, Hayes RP, Roderick TH: Prevalence of retinitis pigmentosa in Maine. Am J Ophthalmol 1984;97:357–365.
15 Grondahl J: Estimation of prognosis and prevalence of retinitis pigmentosa and Usher syndrome in Norway. Clin Genet 1987;31:255–264.
16 Haim M, Holm NV, Rosenberg T: Prevalence of retinitis pigmentosa and allied disorders in Denmark. I. Main results. Acta Ophthalmol (Copenh) 1992;70:178–186.
17 Fishman GA: Retinitis pigmentosa. Genetic percentages. Arch Ophthalmol 1978;96:822–826.
18 Jay M: On the heredity of retinitis pigmentosa. Br J Ophthalmol 1982;66:405–416.
19 Hu DN: Genetic aspects of retinitis pigmentosa in China. Am J Med Genet 1982;12:51–56.
20 Haim M: Retinitis pigmentosa: Problems associated with genetic classification. Clin Genet 1993;44:62–70.
21 Haim M: Epidemiology of retinitis pigmentosa in Denmark. Acta Ophthalmol Scand Suppl 2002;233;1–34.
22 Dryja TP, McGee TL, Reichel E, Hahn LB, Cowley GS, Yandell DW, Sandberg MA, Berson EL: A point mutation of the rhodopsin gene in one form of retinitis pigmentosa. Nature 1990;343:364–366.
23 Phelan JK, Bok D: A brief review of retinitis pigmentosa and the identified retinitis pigmentosa genes. Mol Vis 2000;6:116–124.
24 Bascom RA, Liu L, Heckenlively JR, Stone EM, McInnes RR: Mutation analysis of the ROM1 gene in retinitis pigmentosa. Hum Mol Genet 1995;4:1895–1902.
25 Farrar GJ, Kenna P, Jordan SA, Kumar-Singh R, Humphries MM, Sharp EM, Sheils DM, Humphries P: A three-basepair deletion in the peripherin-RDS gene in one form of retinitis pigmentosa. Nature 1991;354:478–480.
26 Liu X, Vansant G, Udovichenko IP, Wolfrum U, Williams DS: Myosin VIIa, the product of the Usher 1B syndrome gene, is concentrated in the connecting cilia of photoreceptor cells. Cell Motil Cytoskeleton 1997;37:240–252.
27 Sung CH, Tai AW: Rhodopsin trafficking and its role in retinal dystrophies. Int Rev Cytol 2000;195:215–267.
28 Roepman R, Bernoud-Hubac N, Schick DE, Maugeri A, Berger W, Ropers HH, Cremers FP, Fereira PA: The retinitis pigmentosa GTPase regulator (RPGR) interacts with novel transport-like proteins in the outer segments of rod photoreceptors. Hum Mol Genet 2000;9:2095–2105.
29 Freund CL, Gregory-Evans CY, Furukawa T, Papaioannou M, Looser J, Ploder L, Bellingham J, Ng D, Herbrick JA, Duncan A, Scherer SW, Tsui LC, Loutradis-Anagnostou A, Jacobson SG, Cepko CL, Bhattacharya SS, McInnes RR: Cone-rod dystrophy due to mutations in a novel photoreceptor-specific homeobox gene (CRX) essential for maintenance of the photoreceptor. Cell 1997;91:543–553.
30 Swain PK, Chen S, Wang QL, Affatigato LM, Coats CL, Brady KD et al: Mutations in the cone-rod homeobox gene are associated with the cone-rod dystrophy photoreceptor degeneration. Neuron 1997;19:1329–1336.
31 Bessant DA, Payne AM, Mitton KP, Wang QL, Swain PK, Plant C, Bird AC, Zack DJ, Swaroop A, Bhattacharya SS: A mutation in NRL is associated with autosomal dominant retinitis pigmentosa. Nat Genet 1999;21:355–356.
32 Vithana EN, Abu-Safieh L, Allen MJ, Carey A, Papaioannou M, Chakarova C, Al Maghtheh M, Ebenezer MD, Willis C, Moore AT, Bird AC, Hunt DM, Bhattacharya SS: A human homolog of yeast pre-mRNA splicing gene, PRP31, underlies autosomal dominant retinitis pigmentosa on chromosome 19q13.4 (RP11). Mol Cell 2001;8:375–381.
33 Chakarova CF, Hims MM, Bolz H, Abu-Safieh L, Patel RJ, Papaioannou MG, Inglehearn CF, Keen TJ, Willis C, Moore AT, Rosenberg T, Webster AR, Bird AC, Gal A, Hunt D, Vithana EN, Bhattacharya SS: Mutations in HPRP3, a third member of pre-mRNA splicing factor genes, implicated in autosomal dominant retinitis pigmentosa. Hum Mol Genet 2002;11:87–92.
34 Retinal information network: *http.//www.sph.uth.tmc.edu/RetNet 2003*
35 Rosenfeld PJ, Cowley GS, McGee TL, Sandberg MA, Berson EL, Dryja TP: A null mutation in the rhodopsin gene causes rod photoreceptor dysfunction and autosomal recessive retinitis pigmentosa. Nat Genet 1992;1:209–213.

36 Dryja TP: Molecular genetics of Oguchi disease, fundus albipunctatus, and other forms of stationary night blindness: LVII Edward Jackson Memorial Lecture. Am J Ophthalmol 2000;130:547–563.
37 Al Jandal N, Farrar GJ, Kiang AS, Humphries MM, Bannon N, Findlay JB, Humphries P, Kenna PF: A novel mutation within the rhodopsin gene (Thr-94-Ile) causing autosomal dominant congenital stationary night blindness. Hum Mutat 1999;13:75–81.
38 McLaughlin ME, Sandberg MA, Berson EL, Dryja TP: Recessive mutations in the gene encoding the β-subunit of rod phosphodiesterase in patients with retinitis pigmentosa. Nat Genet 1993;4:130–134.
39 Veske A, Orth U, Rüther K, Zrenner E, Rosenberg T, Baehr W, Gal A, et al: Mutations in the gene for the β-subunit of rod photoreceptor CGMP-specific phophodiesterase (PDEB) in patients with retinal dystrophies and dysfunctions; in Anderson RE (ed): Degenerative Diseases of the Retina. New York, Plenum Press, 1995, pp 313–322.
40 Rosenberg T, Schwahn U, Feil S, Berger W: Genotype-phenotype correlation in X-linked retinitis pigmentosa (RP2). Ophthal Genet 1999;20:161–172.
41 Beaufrere L, Claustres M, Tuffery S: No missense mutation in choroideremia patients analyzed to date. Ophthalmic Genet 1999;20:89–93.
42 Donnenfeld AE, Packer RJ, Zackai EH, Chee CM, Sellinger B, Emanuel BS: Clinical, cytogenetic and pedigree findings in 18 cases of Aicardi syndrome. Am J Med Genet 1989;32:461–467.
43 Erlandson A, Bjursell C, Stibler H, Kristiansson B, Wahlström J, Martinsson T: Scandinavian CDG-Ia patients: Genotype/phenotype correlation and geographic origin of founder mutations. Hum Genet 2001;108:359–367.
44 Kjærgaard S, Skovby F, Schwartz M: Absence of homozygosity for predominant mutations in PMM2 in Danish patients with carbohydrate-deficient glycoprotein syndrome type 1. Eur J Hum Genet 1998;6:331–336.
45 Schollen E, Kjærgaard S, Legius E, Schwartz M, Matthijs G: Lack of Hardy-Weinberg equilibrium for the most prevalent PMM2 mutation in CDG-Ia (congenital disorders of glycosylation type Ia). Eur J Hum Genet 2000;8:367–371.
46 Rozet JM, Gerber S, Souied E, Ducroq D, Perrault I, Ghazi I, Soubrane G, Coscas G, Dufier JL, Munnich A, Kaplan J: The ABCR gene: A major disease gene in macular and peripheral retinal degenerations with onset from early childhood to the elderly. Mol Genet Metab 1999;68:310–315.
47 Allikmets R: Further evidence for an association of ABCR alleles with age-related macular degeneration. The International ABCR Screening Consortium. Am J Hum Genet 2000;67:487–491.
48 Kohl S, Giddings I, Besch D, Apfelstedt-Sylla E, Zrenner E, Wissinger B: The role of the peripherin/RDS gene in retinal dystrophies. Acta Anat (Basel) 1998;162:75–84.
49 Klintworth GK: Advances in the molecular genetics of corneal dystrophies. Am J Ophthalmol 1999;128:747–754.
50 Prosser J, van Heyningen V: PAX6 mutations reviewed. Hum Mutat 1998;11:93–108.
51 Hirose M: The role of Wilms' tumor genes. J Med Invest 1999;46:130–140.
52 Grønskov K, Olsen JH, Sand A, Pedersen W, Carlsen N, Bak Jylling AM, Lyngbye T, Brøndum-Nielsen K, Rosenberg T: Population-based risk estimates of Wilms' tumor in sporadic aniridia. A comprehensive mutation screening procedure of PAX6 identifies 80% of mutations in aniridia. Hum Genet 2001;109:11–18.
53 Grønskov K, Rosenberg T, Sand A, Brøndum-Nielsen K: Mutational analysis of *PAX6*: 16 novel mutations including 5 missense mutations with a mild aniridia phenotype. Eur J Hum Genet 1999;7:274–286.
54 Robinson PN, Godfrey M: The molecular genetics of Marfan syndrome and related microfibrillopathies. J Med Genet 2000;37:9–25.
55 Tiecke F, Katzke S, Booms P, Robinson PN, Neumann L, Godfrey M, Mathews KR, Scheumer M, Hinkel GK, Brenner RE, Hövels-Gürich HH, Hagemeier C, Fuchs J, Skovby F, Rosenberg T: Classic, atypically severe and neonatal Marfan syndrome: Twelve mutations and genotype-phenotype correlations in *FBN1* exons 24–40. Eur J Hum Genet 2001;9:13–21.
56 Wallace DC, Singh G, Lott MT, Hodge JA, Schurr TG, Lezza AM, Elsas LJ, Nikooskelainen EK: Mitochondrial DNA mutation associated with Leber's hereditary optic neuropathy. Science 1988;242:1427–1430.

57 Mackey D, Oostra RJ, Rosenberg T, Nikoskelainen E, Bronte-Stewart J, Poulton J, Harding AE, Govan G, Boljuis PA, Nørby S, Bleeker-Wagemakers EM, Savontaus ML, Chan C, Howell N: Primary pathogenic mtDNA mutations in multigeneration pedigrees with Leber hereditary optic neuropathy. Am J Hum Genet 1996;59:481–485.

58 Howell N: Leber hereditary optic neuropathy: Mitochondrial mutations and degeneration of the optic nerve. Vision Res 1997;37:3495–3507.

59 Brown MD, Torroni A, Reckord CL, Wallace DC: Phylogenetic analysis of Leber's hereditary optic neuropathy mitochondrial DNA's indicates multiple independent occurrences of the common mutations. Hum Mutat 1995;6:311–325.

60 Lamminen T, Huoponen K, Sistonen P, Juvonen V, Lahermo P, Aula P, Nikoskelainen E, Savontaus ML: mtDNA haplotype analysis in Finnish families with mtDNA haplotype analysis in Finnish families with Leber hereditary optic neuroretinopathy. Eur J Hum Genet 1997;5:271–279.

61 Torroni A, Petrozzi M, D'Urbano L, Sellitto D, Zeviani M, Carrara F, Carducci C, Leuzzi V, Carelli V, Barboni P, De Negri A, Scozzari R: Haplotype and phylogenetic analyses suggest that one European-specific mtDNA background plays a role in the expression of Leber hereditary optic neuropathy by increasing the penetrance of the primary mutations 11778 and 14484. Am J Hum Genet 1997;60:1107–1121.

62 Hofmann S, Jaksch M, Bezold R, Mertens S, Aholt S, Paprotta A, Gerbitz KD: Population genetics and disease susceptibility: Characterization of central European haplogroups by mtDNA gene mutations, correlation with D loop variants and association with disease. Hum Mol Genet 1997;6:1835–1846.

63 Chinnery PF, Johnson MA, Wardell TM, Singh-Kler R, Hayes C, Brown DT, Taylor RW, Bindoff LA, Turnbull DM: The epidemiology of pathogenic mitochondrial DNA mutations. Ann Neurol 2000;48:188–193.

64 Kajiwara K, Berson EL, Dryja TP: Digenic retinitis pigmentosa due to mutations at the unlinked peripherin/RDS and ROM1 loci. Science 1994;264:1604–1608.

65 Loewen CJ, Moritz OL, Molday RS: Molecular characterization of peripherin-2 and rom-1 mutants responsible for digenic retinitis pigmentosa. J Biol Chem 2001;276:22388–22396.

66 Katsanis N, Ansley SJ, Badano JL, Eichers ER, Lewis RA, Hoskins BE, Scambler PJ, Davidson WS, Beales PL, Lupski JR: Triallelic inheritance in Bardet-Biedl syndrome, a Mendelian recessive disorder. Science 2001;293:2256–2259.

67 Norio R, Nevanlinna HR, Perheentupa J: Hereditary diseases in Finland: Rare flora in rare soil. Ann Clin Res 1973;5:109–141.

68 Eriksson AW, Forsius HR, Nevanlinna HR, Workman PL, Norio RK: Population Structure and Genetic Disorders. London, Academic Press, 1980.

69 De la Chapelle A: Disease gene mapping in isolated human populations: The example of Finland. J Med Genet 1993;30:857–865.

70 Cunier F: Histoire d'une héméralopie héréditaire depuis deux siècles dans une famille de la commune de Vendémian, près Montpellier. Ann Soc Méd Gand 1838;383–395.

71 Nettleship E: A history of congenital stationary night blindness in nine consecutive generations. Trans Ophthalmol Soc UK 1907;27:269–293.

72 Meretoja J: Familial systemic paramyloidosis with lattice dystrophy of the cornea, progressive cranial neuropathy, skin changes and various internal symptoms. A previously unrecognized heritable syndrome. Ann Clin Res 1969;1:314–324.

73 Fossdal R, Magnusson L, Weber JL, Jensson O: Mapping the locus of atrophia areata, a helicoid peripapillary chorioretinal degeneration with autosomal dominant inheritance, to chromosome 11p15. Hum Mol Genet 1995;4:479–483.

74 Nordström S, Barkman Y: Hereditary macular degeneration in 246 cases traced to one gene source in central Sweden. Hereditas 1976;84:163–176.

75 Bakall B, Marknell T, Ingvast S, Koisti M, Sandgren O, Li W, Bergen AAB, Andréasson S, Rosenberg T, Petrutkin K, Wadelius C: The mutation spectrum of the bestrophin protein – Functional implications. Hum Genet 1999;104:383–389.

76 Rosenberg T, Haim M, Piczenik Y, Simonsen SE: Autosomal dominant stationary night-blindness. A large family rediscovered. Acta Ophthalmol (Copenh) 1991;69:694–702.

77 Gal A, Orth U, Schwinger E, Rosenberg T: Heterozygous missense mutation in the rod cGMP phosphodiesterase β-subunit gene in autosomal dominant stationary night blindness. Nat Genet 1994;7:64–68.
78 Eiberg H, Marner E, Rosenberg T, Mohr J: Marner's cataract assigned to chromosome 16: Linkage to haptoglobin. Clin Genet 1988;34:272–275.
79 Eiberg H, Lund AM, Warburg M, Rosenberg T: Assignment of congenital cataract Volkmann type (CCV) to chromosome 1p36. Hum Genet 1995;96:33–38.
80 Wijesuriya SD, Evans K, Jay MR, Davison C, Weber BH, Bird AC, Bhattacharya SS, Gregory CY: Sorsby's fundus dystrophy in the British Isles: Demonstration of a striking founder effect by microsatellite-generated haplotypes. Genome Res 1996;6:92–101.
81 Bessant DA, Payne AM, Plant C, Bird AC, Swaroop A, Bhattacharya SS: NRL S50T mutation and the importance of 'founder effects' in inherited retinal dystrophies. Eur J Hum Genet 2000;8:783–787.
82 Small KW, Hermsen V, Gurney N, Fetkenhour CL, Folk JC: North Carolina macular dystrophy and central areolar pigment epithelial dystrophy. One family, one disease. Arch Ophthalmol 1991;110:515–518.
83 Small KW, Udar N, Yelchits S, Klein R, Garcia C, Gallardo G, Puech B, Puech V, Saperstein D, Lim J, Haller J, Flaxel C, Kelsell R, Hunt D, Evans K, Lennon F, Pericak-Vance M: North Carolina macular dystrophy (MCDR1) locus: A fine resolution genetic map and haplotype analysis. Mol Vis 1999;5:38 *(http.molvis.org/molvis/vol5/p38)*.
84 Votruba M, Moore AT, Bhattacharya SS: Demonstration of a founder effect and fine mapping of dominant optic atrophy locus on 3q28-qter by linkage disequilibrium method: A study of 38 British Isles pedigrees. Hum Genet 1998;102:79–86.
85 Toomes C, Marchbank NJ, Mackey DA, Craig JE, Newbury-Ecob RA, Bennett CP, Vize CJ, Desai SP, Black GC, Patel N, Teimory M, Markham AF, Inglehearn CF, Churchill AJ: Spectrum, frequency and penetrance of OPA1 mutations in dominant optic atrophy. Hum Mol Genet 2001;10:1369–1378.
86 Thiselton DL, Alexander CL, Morris AL, Brooks SL, Rosenberg TL, Eiberg HL, Kjer BL, Kjer PL, Bhattacharya SS, Votruba MS: A frameshift mutation in exon 28 of the OPA1 gene explains the high prevalence of dominant optic atrophy in the Danish population: Evidence for a founder effect. Hum Genet 2001;109:498–502.
87 Boycott KM, Pearce WG, Bech-Hansen NT: Clinical variability among patients with incomplete X-linked congenital stationary night blindness and a founder mutation in *CAGNA1F*. Can J Ophthalmol 2000;35:204–213.
88 Sankila EM, de la Chapelle A, Karna J, Forsius H, Frants R, Eriksson A: Choroideremia: Close linkage to DXYS1 and DXYS12 demonstrated by segregation analysis and historical-genealogical evidence. Clin Genet 1987;31:315–322.
89 Sankila EM, Tolvanen R, van den Hurk JA, Cremers FP, de la Chapelle A: Aberrant splicing of the CHM gene is a significant cause of choroideremia. Nat Genet 1992;1:109–113.
90 Huopaniemi L, Rantala A, Tahvanainen E, de la Chapelle A, Alitalo T: Linkage disequilibrium and physical mapping of X-linked juvenile retinoschisis. Am J Hum Genet 1997;60:1139–1149.
91 The Retinoschisis Consortium: Functional implications of the spectrum of mutations found in 234 cases with X-linked juvenile retinoschisis. Hum Mol Genet 1998;7:1185–1192.
92 Huopaniemi L, Tyynismaa H, Rantala A, Rosenberg T, Alitalo T: Characterization of two unusual RS1 gene deletions segregating in Danish retinoschisis families. Hum Mutat 2000;16:307–314.
93 Bonne-Tamir B, Korostishevsky M, Kalinsky H, Seroussi E, Beker R, Weiss S, Godel V: Genetic mapping of the gene for Usher syndrome: Linkage analysis in a large Samaritan kindred. Genomics 1994;20:36–42.
94 Bonne-Tamir B, Nystuen A, Seroussi E, Kalinsky H, Kwitek-Black AE, Korostishevsky M, Adato A, Sheffield VC: Usher syndrome in the Samaritans: Strengths and limitations of using inbred isolated populations to identify genes causing recessive disorders. Am J Phys Anthropol 1997;104:193–200.
95 Boughman JA, Vernon M, Shaver KA: Usher syndrome: Definition and estimate of prevalence from two high-risk populations. J Chronic Dis 1983;36:595–603.

96 Savas S, Frishhertz B, Peliaz MZ, Batzer MA, Deininger PL, Keats BJB: The *USH1C* 216G→A mutation and the 9-repeat VNTR(t,t) allele are in complete linkage disequilibrium in the Acadia population. Hum Genet 2000;110:95–97.

97 Sacks O: The Island of the Color Blind. New York, Vintage Books, 1998.

98 Sundin OH, Yang JM, Li Y, Zhu D, Hurd JN, Mitchell TN, Silva ED, Maumenee IH: Genetic basis of total colour blindness among the Pingelapese islanders. Nat Genet 2000;25:289–293.

99 Franceschetti A, Jaeger W, Klein D, Ohrt V, Rickli H: Étude patho-physiologique et génétique de la grande famille d'achromates de l'île de Fur (Danemark). XVIII Concilium Ophthalmol Belg 1958;ii:1582–1588.

100 Gershoni-Baruch R, Rosenmann A, Droetto S, Holmes S, Tripathi RK, Spritz RA: Mutations of the tyrosinase gene in patients with oculocutaneous albinism from various ethnic groups in Israel. Am J Hum Genet 1994;54:586–594.

101 Stevens G, Ramsay M, Jenkins T: Oculocutaneous albinism (OCA2) in sub-Saharan Africa: Distribution of the common 2.7-kb P gene deletion mutation. Hum Genet 1997;99:523–527.

102 Lund PM, Puri N, Durham-Pierre D, King RA, Brilliant MH: Oculocutaneous albinism in an isolated Tonga community in Zimbabwe. J Med Genet 1997;34:733–735.

103 Dreyer B, Tranebjaerg L, Brox V, Rosenberg T, Möller C, Beneyto M, Weston MD, Kimberling WJ, Nilssen Ø: A common ancestral origin of the frequent and widespread 2299delG *USH2A* mutation. Am J Hum Genet 2001;69:228–234.

Thomas Rosenberg, MD
Gordon Norrie Centre for Genetic Eye Diseases,
National Eye Clinic for the Visually Impaired,
1 Rymarksvej, DK–2900 Hellerup (Denmark)
Tel. +45 3945 2400, Fax +45 3945 2420, E-Mail roseeye@visaid.dk

Wissinger B, Kohl S, Langenbeck U (eds): Genetics in Ophthalmology.
Dev Ophthalmol. Basel, Karger, 2003, vol 37, pp 34–49

Interactions of Genes and Environment in Myopia

Marita Feldkämper, Frank Schaeffel

Section of Neurobiology of the Eye University Eye Hospital Department II, Tübingen, Germany

Abstract

Myopia is a condition in which the eye is too long for the focal length of cornea and lens, and the plane of sharp focus ends up in front of the retina. Given that the growth of the length of the eye is normally controlled with extreme precision by an image-processing feedback mechanism in the retina, myopia can either be the result of inappropriate visual stimulation, genetically determined changes in the gain or offset of the feedback loops or of inappropriate responses of the target tissues. There is no doubt that an environmental component is involved and extended near work appears to be the major risk factor. However, there is also no doubt that myopia is inherited since myopic parents are much more likely to have myopic children, and myopia is far more frequent in Asian populations than in the USA or Europe, even if groups are compared that have performed similar amounts of near work. A number of systemic or ophthalmic diseases are associated with myopia, indicating that metabolic conditions may interfere either with the gains of the feedback loops or the responses of the target tissue, the sclera. Since there is still no therapy against myopia development, research is directed toward the identification of genes that control the axial elongation of the eye.

What Is Myopia?

In a normal-sighted ('emmetropic') eye, the image of distant objects is focused on the retina when accommodation is relaxed. With this condition, the full accommodation amplitude is available to focus on to close objects. To achieve emmetropia and, accordingly, optimal visual acuity over a wide range of viewing distances, the length of the eye must be precisely matched to the focal length of the cornea and the lens. The required precision is about 0.1 mm. The observation that many eyes are emmetropic argues against the assumption that a mismatch between both variables results from random variation of growth. It is rather likely that the feedback loop controlling the match between both variables has either inappropriate input (environmental factors) or that elements of the loop do not function properly or have to high gain (genetic factors).

In myopia, the eye is too long and the image is in front of the retina during distant vision. Accordingly, accommodation is not needed to focus close objects (the only advantage is then that presbyopic older subjects, whose accommodation is vanished, can read without reading glasses). Myopia can be readily corrected by negative lenses. However, the elongation of the globe has side effects like increased risk of retinal detachment (×10), glaucoma (×2–5), or chorioretinal degeneration (×10). In these cases, myopia also increases the risk of blindness.

Prevalence of Myopia Around the World and Its Association with Environmental Factors

Examples of the frequency of myopia in different recent studies is shown in table 1. It can be seen that an average of 30% of myopia over the world will not be overestimated.

While there is a clear association between the risk of becoming myopic and the number of myopic parents (discussed in more detail below: Evidence for Genetic Control of Myopia), there is also evidence for a role of environmental factors. For example, Wu and Edwards [1] have compared the prevalence of myopia in 3 generations of Chinese (total number of subjects studied: 21,137) and have found that 5.8% of the grandparents were myopic, 20.8% of the parents' generation, and 26.2% of the children's' generation. Given that the children were 7–17 years old, a further increase is expected in the youngest generation. Similar results were obtained from the Framingham Offspring Study [2] in which it was found that the incidence of myopia increased from 20% for 65 years and older to 60% at the ages of 23–34 years. The increase in the incidence of myopia cannot be attributed to changes in the genetical background but must rather result from environmental influences. A few examples of recent studies (among many others) confirm this conclusion: (1) Zylberman et al. [3] found myopia in 27.4% of the Jewish male students in a public school and 81.3% in an orthodox school with particularly high reading demands. (2) Pärssinnen and Lyyra [4] found that myopia progression in schoolchildren in Finland was associated with the number of hours of near work as well as with the reading distance. (3) Hepsen et al. [5] found that myopia progression was higher in students than in laboratory workers. (4) Saw et al. [6] found a higher likelihood that 7- to 9-year-old Chinese children develop myopia when they did more reading or close-up work and that professional work increased the risk of myopia in Singapore woman. (5) Tan et al. [7]

Table 1. Prevalence of myopia (<−0.5 D spherical equivalent)

Country	Group studied	Age	n	Prevalence, %	Group (first author)
China	Schoolchildren	15	5,884	55	Zhao, 2000
Nepal	Schoolchildren	15	5,067	3	Pockhard, 2000
Chile	Schoolchildren	15	5,303	14.7	Maul, 2000
Denmark	Students	26	294	53.9	Fledelius, 2000
Singapore	Students	20	1,232	>90	Wong, 2000
Singapore	Rural	7	132	3.9	Zhang, 2000
Singapore	Urban	7	104	9.1	Zhang, 2000
Singapore	City	7	146	12.3	Zhang, 2000
South India	All	>15	2,321	19.4	Dandona, 2000
Hong Kong	Schoolchildren	10	142	63	Lam, 2000
Hong Kong	Schoolchildren	12	83	59	Edwards, 1999
Australia	All	40–49	5,740	17	Wensor, 2000
Oman	Schoolchildren	12	6,292	5.16	Lithander, 1999
Sweden	Schoolchildren	12	1,045	49.7	Villarreal, 2000
USA	Schoolchildren	12	6,000	15	Zadnik, 1993
Taiwan	Students	17	11,178	84	Lin, 1999
Tibet	Schoolchildren	12	555	21.7	Garner, 1999
Nepal	Schoolchildren	12	270	2.9	Garner, 1999
Norway	Students	20.6	224	65	Kinge, 1999
Japan	Students	17	346	66	Matsumara, 1999
Israel	All		312,149	16.3	Rosner, 1999
Greece	Students	15–18	220,000	36.8 (<−0.25 D)	Mavracanas, 2000

found that myopia progression in children in Singapore varies over the academic year, increasing preferentially after the final school examinations.

It is interesting to note that the nature of the environmental factors that promote myopia is difficult to define, despite extensive research in animal models (see chapter 3). It is likely that it is associated with reading but what kind of visual experience during reading triggers myopia is still not clear. Some authors [6] state that their study 'does not unambiguously resolve whether near work is a risk factor for the development of myopia or a surrogate for other environmental or genetic factors'. They [8] have also reviewed the interactions of genes and environment in myopia. Both insufficient accommodation during reading ('lack of accommodation') which places the focused image slightly behind the retina and too much accommodation, assumed to place mechanical pressure on the walls of the globe and enlarge it, have been proposed to promote myopia. In addition, 'deprivation' of the retina of fractions of the spatial frequency spectrum has been proposed to occur during reading [9]. It is known from animal models that degradation of the retinal image promotes myopia development (see chapter 4).

Evidence for Visual Control of Eye Growth in Animal Models

Experimental Manipulations of Visual Experience

Although there are early studies in monkeys suggesting that restriction of the viewing distance can induce myopia [10], a role of visual experience in myopia was fully accepted after the study by Wiesel and Raviola [11] which was published in *Nature*. These authors had found that unilateral lid fusion produced extensive axial myopia in monkeys in the closed eyes, based on a visual effect. Wallman et al. [12] described that covering the eyes with frosted occluders produced extreme myopia in chickens and showed that the retina processes the projected image and controls the growth of the underlying sclera [9]. Later, it was found that the retina cannot only determine whether the image is in focus or poor, but also on which side of the photoreceptor layer the focal plane it is [13]. Treating animal models with lenses produces predictable refractive errors, myopia and longer eyes in the case of negative lenses (which place the image behind the retina) and hyperopia and shorter eyes in the case of positive lenses (which place the image in front of the retina). This observation was made in chickens [14] as well as in tree shrews [15], marmosets [16] and rhesus monkeys [17]. While it provides convincing evidence that the growing eye can use visual cues to control its refractive development, it remains unclear how these findings can be extrapolated to human myopia. There is evidence that humans accommodate insufficiently during reading which would place the plane of focus behind the retina just as the negative lenses in animal models. However, daily interruption of occluder treatment for only a few minutes reduces myopia very powerfully [18] and humans interrupt reading probably much longer every day. A very recent study could show that adolescent monkeys, working on visual tasks on a computer monitor, become the more myopic the closer the monitors were positioned [19]. These results confirm the assumed association between near work and myopia. They open the field for more specific studies on the nature of the underlying visual stimuli.

Genetical Control of Experimental Myopia in Animal Models: Gain Problems and Naturally Occurring Myopia in Animal Models

A major problem of myopia studies in animal models is that they would probably not have developed myopia without the artificial manipulations of their visual environment. Since human subjects doing near work may either become myopic or not, unknown genetic factors must make the difference. One possibility is that the ‘gain’ of the ‘feedback loop’ that controls eye growth from retinal image processing is genetically determined [20]. The authors had found that deprivation myopia is very variable among individuals but still symmetrical in both eyes of individual chickens, despite that myopia is independently regulated. Troilo [pers. commun.] induced myopia twice, with a period of recovery in between, and found that animals that had developed much myopia the first time, responded similarly also the second time. There are striking differences in the susceptibility of different chicken strains to experimental myopia which must be genetical [21, 22]. It would be very useful to have animal models available that develop myopia with normal visual experience. This would, in principle, make it possible to identify underlying gene loci. However, up to now there were no naturally occurring myopias in animal models found, other than in some dog strains [i.e. 23]. However, they have not yet been experimentally studied.

Evidence for Genetic Control of Myopia

Parent-Offspring Relationships

There is abundant evidence that myopic parents are more likely to have more myopic children. The apparent heritability was variable among studies (which is understandable, given the different genetic backgrounds of the studied samples and the different ages of the children at the time of refraction). In the Orinda study (USA), 40% of the children were myopic when both parents were myopic, 20–25% if one parent was myopic and 10% if no parent was myopic [24]. The respective values in a study by Sorsby and Benjamin [25] were 0.07/0.25/0.25–0.4, in a study by Wu and Edwards [1] 0.22/0.31/0.46, in a study by Pacella et al. [26] approximately 0.3/0.25/0.57. There are also studies which did not find a relationship between myopia in parents and offspring (in Hong Kong Chinese children, probability with one myopic parent 0.55, with two myopic parents 0.6 [27]). In this case, it was proposed that the genotype may not have been expressed in the parents. Based on tests with three models, using the data of the Orinda study, Mutti [pers. commun.] refused the hypothesis that a genetically determined gain is responsible for the variability of experimental myopia. He claimed that only 10–12% of myopia is environmentally determined and the effect of genetic factors is dominant. In these studies it was also examined whether the reading habits may have been inherited from the parents rather than the myopia itself [28].

Twin Studies

Twin studies have confirmed that myopia has a major genetic component. In a study by Hammond et al. [29] on 226 monozygotic adult twins in the UK (who share 100% of the genes) and 280 dizygotic adult twins (who share 50%

of the genes), the heritability of refractive errors was determined to between 84 and 86%. This fits to the conclusions of the Orinda study on children (see above: Parent-Offspring Relationships). If myopia and hyperopia were treated as separate traits, the heritability was even higher (90% for myopia). Teikari et al. [30] conducted a number of twin studies in Finland. They showed that the mean difference in refraction between monozygotic twins was 1.19 D in the right eyes and 1.15 D in the left. In dizygotic pairs, these numbers were 2.34 and 2.47, respectively (more variance in dizygotic pairs, $p < 0.001$). In this study, heritability of myopia was found to be 0.58 (higher in males (0.74) than in females (0.61)) [30]. In a previous study on 3,676 monozygotic twin pairs and on 8,109 dizygotic twin pairs, the heritability was 0.62 among males and 0.98 among females. Lower heritability was found in studies on 90 monozygotic twins and 36 pairs of like-sex dizygotic twin pairs by Lin and Chen [31] on Chinese schoolchildren (0.65 in monozygotic twins and 0.46 in dizygotic twins). In contrast, in a study on a smaller sample of 5-year twins studied by Angi et al. [32], the heritability was only 0.08–0.14. There are also some striking observations in twin studies: in one case, both of a monozygotic twin pair, aged 64, had 20 D of myopia only in their left eyes [33]. In another case, anisometropia was a mirror image in both monozygotic twins in two cases [34].

Predictability of Myopia in Children

Myopia development can be predicted in children with some confidence. Although it is possible that their studying habits are inherited and determine the development of myopia only indirectly, it is also likely that genes control eye growth directly. The latter assumption is supported by a study of Zadnik et al. [24] who claimed that children with two myopic parents had larger eyes already before they developed myopia. They also described that these children had less hyperopic start-up refractions than children with only one or no myopic parent. That early refraction history predicts future myopia has been recognized several times: Zadnik et al. [35] state that the probability of becoming myopic within 5 years in 8-year-old children is about 39% if the spherical equivalent refraction is zero, 9% if it is +0.5 D or more hyperopic and 4% if it is +0.75 D or more hyperopic. Specificity was 73.3% and sensitivity 86.7%. Another group [26], studying infants, arrived at a similar conclusion: 'children who had refractions in the lower half of the distribution at 6 to 12 months of age were 4.33 times more likely to develop myopia'. Among the first who arrived at this conclusion were Howland et al. [36].

Different Gains in Different Populations

That genetic factors determine the development of myopia can also be inferred from comparative studies between different countries. Several authors have found that myopia progression is much faster in the Asian countries. Grice et al. [37] compared myopia progression in children in Boston (n = 79) and Singapore (n = 92). The progression rates were 0.37 D/year in Boston and 1.16 D/year in Singapore at the age of 12 years which represents a factor of 3 times faster. Myopia progression in children from Hong Kong was about 1 D/year at the age of 6 years and declined to about 0.4 D/year at the age of 15 and stated that the progression is much faster than in European or American children [38]. Similar observations were made by others [6] who also confirmed that the prominent difference in myopia progression in the two populations cannot be attributed to differences in sociodemographic associations.

A Selection of Systemic or Ophthalmic Diseases Known To Be Associated with Myopia, and Their Gene Loci

A wide variety of hereditary systemic and ocular disorders is associated with myopia and, in these cases, myopia follows the mode of inheritance of the accompanying disease. Chromosomal abnormalities as well as pre- and perinatal disorders can produce myopia, often of an unusually high degree. Myopia has been reported in association with numerous syndromes and diseases. Examples are presented below.

Albinism

Albinism represents a group of inherited abnormalities of melanin synthesis characterized by a congenital reduction or absence of melanin pigment. Hypopigmentation gives rise to specific developmental changes in the visual system. Oculocutaneous albinism type 1 and 2 (OMIM: 203100, 203200) involves the skin and hair, as well as the visual system, including the eye and the optic nerves. It exhibits an autosomal recessive mode of inheritance. Ocular albinism is accompanied by a reduction of the pigment content in the retinal pigment epithelium of the eye. It exhibits an X-linked (OMIM: 300500, 300600) or autosomal recessive mode of inheritance (OMIM: 203310). In both forms of albinism, myopia, often of high degree, may be encountered. Recent evidence has shown that albinism is a heterogeneous genetic disorder caused by mutations in several different genes [39]. At present, a range of genetic loci responsible for human albinism have been mapped (gene locus and assignment: OA1 located on Xp22.3–22.2, OA2 on Xp11.4–11.23, OA3 on 6q13–15, OCA1 on 11q14.3 and OCA2 on 15q11.2–12). The genes have been isolated and pathological mutations identified (tyrosinase, OA1, P-gene – the human homologue of the mouse pink-eyed dilution gene).

Cohen Syndrome

The Cohen syndrome (OMIM: 216550) is a rare disorder characterized by infantile hypotonia, childhood obesity and numerous dysmorphic features. Progressive myopia and retinochoroidal dystrophy are found in a large proportion of the patients [40]. The Cohen syndrome gene was assigned to gene map locus 8q22–q23.

Down Syndrome

An increase in the frequency of refractive errors in individuals with Down's syndrome (OMIM: 190685), caused by triplicate state of all or a critical portion of chromosome 21, has been documented by many authors [e.g. 41]. Different from normally developing children, refractive errors increase with development among children with Down's syndrome. Prevalence of myopia in mongolism is about 12–40% and prevalence of hyperopia is about 20–25%. Myopia is severe in more than half of myopic patients.

Ehlers-Danlos Syndrome (EDS)

EDS is a clinically and genetically heterogeneous connective tissue disorder. Following the identification of specific mutations in the genes encoding collagen types I, III and V, as well as several collagen-processing enzymes, the Villefranche classification of EDS was collapsed into six distinct clinical syndromes (table 2) emphasizing the molecular basis of each form [42]. Myopia is common in type I (75%), type II (50%) and type IV (65%).

Marfan's Syndrome

Marfan's syndrome (OMIM: 154780) is an autosomal, dominantly inherited disorder of connective tissue in which cardiovascular, skeletal, ocular and other abnormalities may be present to a highly variable degree [43]. The clinically apparent features are the result of a weakening of the supporting tissues, due to defects in fibrillin-1, a glycoprotein and a principal component of the extracellular matrix microfibril. The gene for fibrillin-1 (FBN1) is located on chromosome 15q21.1. More than 200 mutations in FBN1 have been described. The phenotype is highly variable due to varying genotype expression [44]. Prevalence of myopia, often at high degree, varies among studies between 28 and 83% [45]. The Marfan-like disorder (OMIM: 154705) which is localized in 3p24.2–p25 and which was described in a large French family, lacks ocular abnormalities.

Marshall Syndrome

Marshall syndrome (OMIM: 15478) is a dominant disorder characterized by craniofacial and skeletal abnormalities, sensorineural hearing loss, high

Table 2. Villefranche classification of Ehlers-Danlos syndrome

Type	Inheritance	OMIM	Gene defects	Gene map locus
I/II	Autosomal dominant	130010 130000	COL5A1, COL5A2, COL1A1	17q21.31–q22; 9q34.2–q34.2; 2q31
III	Autosomal dominant	120180	COL3A1	2q31
IV	Autosomal dominant	130050	COL3A1	2q31
VI	Autosomal recessive	225400	Lysin hydroxylase	1p36.3–p36.2
VIIa,b	Autosomal dominant	130060	COL1A1, COL1A2	17q21.31–q22; 7q22.1
VIIc	Autosomal recessive	225410	Procollagen N-peptidase	5q23

myopia and cataracts. It is associated with splicing mutations in COL11A1 on chromosome 1p21, demonstrating allelism of Marshall syndrome with a subset of Stickler syndrome families associated with COL11A1 mutations. Myopia is the most common eye problem in Marshall syndrome also, but cataracts occur more frequently and detached retina less frequently than in Stickler's syndrome. The distinctness of the Marshall and the Stickler syndrome is strongly supported by the work of Ayme and Preus [46]. Genotypic-phenotypic comparisons revealed an association between the Marshall syndrome phenotype and the splicing mutations of the 54-bp exons in the C-terminal region of the COL11A1 gene. Null-allele mutations in the COL2A1 gene led to the typical phenotype of Stickler's syndrome. Some patients, however, exhibited phenotypes of both Marshall and Stickler's syndromes.

Stickler's Syndrome

Stickler's syndrome is one of the most common hereditary disorders that affect the body's collagen. It was first discovered and documented by Stickler et al. [47] who associated it with severe myopia. Prevalence for myopia is 75–90% [48]. Stickler's syndrome is an autosomal dominant genetic progressive condition. Not all symptoms are present at birth and may appear as development occurs. Collagen is the major protein found in the body's connective tissue, cartilage and bone. Current research has detected specific genes which are responsible for the body's collagen synthesis and breakdown. They have been proposed as a causal factor for Stickler's syndrome. A mutation in COL2A1 (type II collagen), which is the most abundant collagen found in cartilage and vitreous, causes Stickler's syndrome in 75% of the population diagnosed with the disorder. It has been classified as type I Stickler syndrome (OMIM: 108300). The gene has been linked to chromosome12 (12q13.11–q13.2) in many families. Symptoms include joint, sensorineural hearing loss, ocular and craniofacial abnormalities. COL11A1 (type XI collagen) has been linked to chromosome 1p21 in some families, noted as causing type II Stickler syndrome (OMIM: 604841). The symptoms presented are similar to type I Stickler syndrome. In contrast, mutations of COL11A2 (type XI collagen), which has been linked to chromosome 6p21.3, cause Stickler's syndrome without ocular abnormalities (OMIM: 184840).

Myopia Associated with Ocular Diseases

Myopia associated with ocular disease is less numerous than systemic disorders, although a substantial number of ocular diseases are accompanied by myopia. These will not be described in detail in this chapter. The Xp11 region of the human genome is a gene-rich region of particular interest because it includes several disease genes such as X-linked congenital stationary night blindness (CSNB1). Night blindness is a symptom of several chorioretinal degenerations. There is an autosomal dominant and an X-linked form. The X-linked form is distinguished from the autosomal form by its association of myopia.

Candidate Genes Studied in Myopic Subjects

Until now, the only 'myopia genes' that have been identified are those associated with systemic connective disorders like Marfan's and Stickler's syndrome. It is known that these particular genes are not responsible for causing high myopia in the general population. Refractive error occurs as a continuum across the population and as such are likely to be multifactorial in origin with a complex mode of inheritance [49]. The genetics of myopia is complex and it is rarely possible to find families showing a clear-cut monofactorial (Mendelian) inheritance pattern. Instead, it is thought that a number of interacting genes determine whether an individual develops myopia, and also the final severity of the trait. Karlsson [50] proposed that inheritance is probably autosomal recessive. Although autosomal recessive inheritance has been suggested by others, autosomal dominant myopia has also been reported [e.g. 51].

As part of a 24-year longitudinal study of refractive error and visual development, a three-generation family affected by juvenile-onset myopia has been identified. After an attempt to locate the gene(s) which may influence myopia susceptibility, Grice et al. [52] concluded that their results were not consistent with a simple Mendelian inheritance and suggest that juvenile myopia is likely to be inherited as a complex trait. Their study showed that the gene responsible for Stickler's syndrome is not a cause of juvenile-onset myopia. Moreover, the myopic trait affecting this pedigree is not caused by the genes located at the well-known myopia loci MYP2 and MYP3 [53]. Allelic association between the TIGR (trabecular meshwork-induced glucocorticoid response) gene and severe sporadic myopia was investigated by Wu et al. [54]. No significant linkage of TIGR to severe myopia could be demonstrated by sib-pair analysis. The discovery of an association between myopia and TIGR through transmission disequilibrium test suggests that TIGR is involved in the development of severe sporadic myopia.

Recently, genes expressed in a human scleral cDNA library were identified [55] and provide the first comprehensive list of genes expressed in human sclera. Identification of genes expressed preferentially or exclusively in sclera contributes to the understanding of scleral biology, and provides positional candidate genes for scleral disorders such as high myopia.

Linkage Analysis in Pedigrees

Extremes of refractive error, as high myopia (usually classified as myopia more than –6 D), are more likely to exhibit a simple mode of inheritance and the characterization of the inheritance of these may reveal molecular mechanisms relevant to refractive variation in general. Genetic studies in a small group of carefully selected families have identified the chromosomal locations of high myopia genes which appear to produce a direct genetic effect. In this case, individuals who inherit defective copies of either of these two genes invariably go on to develop high myopia, irrespective of environmental influences.

MYOPIA 1

The Bornholm eye disease (OMIM: 310460) described in one family in 1988, consists of X-linked high myopia and reduced electroretinographic flicker responses with abnormal photopic components [56]. The disease has been mapped to chromosome Xq28 (MYP1 gene locus). In a second family, also of Danish descent, affected individuals had mean increased high myopia (–13.18 D), high cylinder (1.81 D), and axial length (28.39 mm) and subnormal photopic electroretinogram results. This kindred phenotypically resembles the one described for the Bornholm eye disease, and supports chromosome Xq27.3–q28 mapping of the MYP1 locus [57]. Phenotypic features of this kindred also overlap with other types of retinal cone dystrophies that map to Xq27 and Xq28. Mutation screening results in this family ruled out matches with those forms of cone dysfunction, however. The Bornholm eye disease and this phenotype may be similar disorders, and may represent a newly described X-linked cone dystrophy.

MYOPIA 2

Recently, loci for autosomal dominant high myopia have been identified on chromosome 18p, MYP2 (OMIM: 160700) being the symbol for this first form of autosomal dominant myopia. Young et al. [58] conducted a genomewide screen for myopia susceptibility loci in families with an autosomal dominant pattern of myopia of more than –6 D. Linkage analysis refined this myopia locus to a 7.6 cM interval between markers D18S59 and D18S1138 on 18p11.31. It was suggested that the gene for 18p11.31-linked high myopia is most proximal to marker D18S52, with a likely interval of 0.8 cM between markers D18S63 and D18S52. With this contraction of the interval size by transmission disequilibrium tests, the authors concluded that their results provided a basis for focused positional cloning and candidate gene analysis at the MYP2 locus. Laminin (a connective tissue protein that helps cells and matrix components to attach to their surrounding environment) is situated in the region of chromosome 18 implicated and is, thus, a likely candidate. The MYP2 locus was moreover recently confirmed independently in a different patient population with severe autosomal dominant myopia in an Italian family.

MYOPIA 3

A second type of autosomal dominant high-grade myopia was identified [59]. The authors excluded genetic linkage to 18p11.31 and demonstrated linkage to 12q21–q23 in a large German/Italian family (OMIM: 603221). The symbol for this second type of autosomal dominant myopia is MYP3. The family had no clinical evidence of connective-tissue abnormalities or glaucoma and markers flanking or intragenic to the genes for the 18p locus, Stickler syndromes type I and II, Marfan's syndrome and juvenile glaucoma showed no linkage to the myopia in this family. The maximum lod score with 2-point linkage analysis was 3.85 at a recombination fraction of 0.0010, for markers D12S1706 and D12S327. Recombination events defined a 30.1 cM interval on 12q21–q23 for this second autosomal myopia gene. Analysis pointed to decorin which maps to 12q23, and lumican which maps to 12q21.3–q22, as candidate genes. These are members of the small interstitial proteoglycan family of proteins that are expressed in the extracellular matrix of various tissues. Both interact with collagen and limit the growth of fibril diameter. Dermatan sulfate proteoglycan-3, which maps to 12q21, is another small interstitial proteoglycan that is expressed in cartilage, as well as in ligaments and the placental tissues. Its presence in sclera has apparently not been demonstrated. Young et al. [60] suggested that fibrillogenesis of the sclera may be affected by mutations in these candidate proteins. New heteroduplex and sequence analysis exclude lumican as the causative gene in 12q21–23-linked high myopia. Moreover, mutational screening of decorin gene in 90 Chinese probands with high myopia revealed that mutations in the decorin gene are not responsible for pathologic high myopia in these families.

Novel Loci

A new locus for autosomal dominant high myopia maps to chromosome 17q21–23 [61]. In order to find new loci implicated in high myopia, Naiglin et al. [62] conducted a genome screen in 21 French and 2 Algerian families following an autosomal dominant mode of inheritance with weak penetrance. This study suggests a novel locus for high myopia on chromosome 7q36, the locus for pigment dispersion syndrome. The implicated region is an 11.7 cM interval extending from S7S798 to the telomeric end of the chromosome. The computational search for genes and/or expressed sequence tags physically mapped between markers D7S798 and the telomere showed numerous unidentified transcripts, mRNAs for an open reading frame, and several genes. None of these appeared to be good candidate genes on the basis of their known function. In addition, there is no evidence of any closely related genes shared by the regions of interest on chromosomes 7q35, 12q21–23, and 18p11.31.

These above-mentioned results confirm the assumption of a genetic heterogeneity of myopia and the identification of the involved genes may provide insight into the pathophysiology of myopia and eye development in the future.

Studies of Candidate Genes in Experimental Myopia

Because only little is known about the critical set of genes that modulate rates and duration of normal eye growth, Williams and Zhou [63] exploited complex trait analysis to map subsets of genes that control normal differences in eye growth among mice. They explored the genetic basis of variation in eye size and specifically mapped genes that modulate eye weight, lens weight and retinal surface area. Their goal was to characterize genes that influence susceptibility and progression of myopia in humans. Eye1 and Eye2 were the first loci shown to control normal variation in eye size in any mammal. The hepatic growth factor gene, a potent mitogen expressed in retina and RPE, is a strong candidate for Eye1, whereas peripherin 2 and a retinoid X receptor (Rxrb) are candidates for Eye2. The human homologue of Eye2 should map to 7q, that of Eye2 to 6p21, 16q13.3, or 21q22.3. Four new Eye loci have been added in 2001.

Conclusions

Myopia is an example of a very frequent ocular disorder in which environmental and genetic influences tightly interact. It was only recently studied using modern molecular genetic techniques and these approaches may finally permit to develop a pharmacological therapy against myopia development.

References

1 Wu MMM, Edwards MH: The effect of having myopic parents: An analysis of myopia in three generations. Optom Vis Sci 1999;76:387–392.
2 Framingham Offspring Eye Study Group: Familial aggregation and prevalence of myopia in the Framingham Offspring Eye Study. Arch Ophthalmol Chic 1996;114:326–332.
3 Zylberman R, Landau D, Berson D: The influence of study habits on myopia in Jewish teenagers. J Pediatr Ophthalmol Strabismus 1993;30:319–322.
4 Pärssinen O, Lyyra AL: Myopia and myopic progression among schoolchildren: A three-year follow-up study. Invest Ophthalmol Vis Sci 1993;34:2794–2802.
5 Hepsen IF, Evereklioglu C, Bayramlar H: The effect of reading and near-work on the development of myopia in emmetropic boys: A prospective, controlled, three-year follow-up study. Vision Res 2001;41:2511–2520.
6 Saw SM, Chua WH, Hong CY, Wu HM, Chan WY, Chia KS, Stone RA, Tan D: Near work in early onset myopia. Invest Ophthalmol Vis Sci 2002;43:332–339.
7 Tan NWH, Saw SM, Lam DSC, Cheng HM, Rajan U, Chew SJ: Temporal variation in myopia progression in Singaporean children within the academic year. Optom Vis Sci 2000;77:465–472.
8 Saw SM, Chua WH, Wu HM, Yap E, Chia KS, Stone RA: Myopia: gene-environment interaction. Ann Acad Med Singapore 2000;29:290–297.
9 Wallman J, Gottlieb MD, Rajaram V, Fugate-Wentzek LA: Local retinal regions control local eye growth and myopia. Science 1987;237:73–77.
10 Young F: The effect of restricted visual space on the primate eye. Am J Ophthalmol 1961;52: 799–806.
11 Wiesel TN, Raviola E: Myopia and eye enlargement after neonatal lid fusion in monkeys. Nature 1977;266:66–68.
12 Wallman J, Turkel J, Trachtman J: Extreme myopia produced by modest change in early visual experience. Science 1978;201:1249–1251.
13 Schaeffel F, Diether S: The growing eye: An autofocus system that works on very poor images. Vision Res 1999;39:1585–1589.
14 Schaeffel F, Glasser A, Howland HC: Accommodation, refractive error and eye growth in chickens. Vision Res 1988;28:639–657.
15 Siegwart J, Norton T: Refractive and ocular changes in tree shrews raised with plus or minus lenses. Invest Ophthalmol Vis Sci 1993;34(suppl):1208.
16 Whatham AR, Judge SJ: Compensatory changes in eye growth and refraction induced by daily wear of soft contact lenses in young marmosets. Vision Res 2001;41:267–273.
17 Hung LF, Crawford MLJ, Smith EL: Spectacle lenses alter growth and refractive state in young monkeys. Nat Med 1995;1:761–765.
18 Smith EL, Hung LF, Kee CS, Qiao Y: Effects of brief periods of unrestricted vision on the development of form-deprivation myopia in monkeys. Invest Ophthalmol Vis Sci 2002;43:291–299.
19 Bradley DV, Smith EL, Harwerth RS, Fernandes A: A near-work-induced myopic shift, without sustained accommodation, in primates. Invest Ophthalmol Vis Sci 2002;43(suppl):194.
20 Schaeffel F, Howland HC: Properties of the feedback loops controlling refractive state in the chicken. Vision Res 1991;31:717–734.
21 Schmid KL, Wildsoet CF: Breed and gender differences in eye growth and form deprivation myopia in chick. J Comp Physiol A 1996;178:551–561.
22 Troilo D, Glasser A, Howland HC: Differences in eye growth and the response to visual deprivation in different strains of chicks. Vision Res 1995;35:1211–1216.
23 Murphy CJ, Zadnik K, Mannis MJ: Myopia and refractive errors in dogs. Invest Ophthalmol Vis Sci 1992;33:2459–2463.
24 Zadnik K, Satariano WA, Mutti DO, Sholtz RI, Adams AJ: The effect of parental history of myopia on children's eye size. JAMA 1994;271:1323–1327.
25 Sorsby A, Benjamin B: Modes of inheritance of errors of refraction. J Med Genet 1973;3: 269–273.

26 Pacella R, McLellan J, Grice K, Bono EA, Wiggs JL, Gwiazda JE: Role of genetic factors in the etiology of juvenile-onset myopia based on a longitudinal study of refractive error. Optom Vis Sci 1999;76:381–386.
27 Edwards MH: Effect of parental myopia on the development of myopia in Hong Kong Chinese. Ophthalmic Physiol Opt 1998;18:477–483.
28 Zadnik K: The Glenn A. Fry Award Lecture (1995). Myopia development in childhood. Optom Vis Sci 1997;74:603–608.
29 Hammond CJ, Snieder H, Gilbert CE, Spector TD: Genes and environment in refractive error: The twin eye study. Invest Ophthalmol Vis Sci 2001;42:1232–1236.
30 Teikari JM, O'Donnel J, Kaprio J, Koskenvuo M: Impact of heredity in myopia. Hum Hered 1991;41:151–156.
31 Lin LL, Chen CJ: Twin study on myopia. Acta Genet Med Gemellol (Roma) 1987;36:535–540.
32 Angi MR, Clementi M, Sardei C, Piatelli E, Bisantis: Heritability of myopic refractive errors in identical and fraternal twins. Graefes Arch Clin Exp Ophthalmol 1993;231:580–585.
33 DeJong PT, Oostra BA, DeFaber JT: High symmetric anisometropia in monozygotic twins. Ophthalmic Pediatr Genet 1993;14:29–32.
34 Okamoto F, Nonoyama T, Hommura S: Mirror image myopic anisometropia in two pairs of monozygotic twins. Ophthalmologica 2001;215:435–438.
35 Zadnik K, Mutti DO, Friedman NE, Qualley PA, Jones LA, Qiu PH, Kim HS, Hsu JC, Moschberger ML: Ocular predictors of onset of juvenile myopia. Invest Ophthalmol Vis Sci 1999;40:1936–1943.
36 Howland HC, Waite S, Peck L: Early focusing history predicts later refractive state: A longitudinal photorefractive study. OSA 1993 Technical Digest Series on Noninvasive Assessment of the Visual System 1993;3:210–213.
37 Grice K, Bauer J, Tham I, Rajan U, Thorn F, Cheng HM, Chew SJ, Gwiazda J: Myopic progression differs between children in Boston and Singapore. Invest Ophthalmol Vis Sci 1997; 38(suppl):4539.
38 Lam CSY, Edwards M, Millodot M, Goh WSH: A two-year longitudinal study of myopia progression and optical component changes among Hong Kong schoolchildren. Optom Vis Sci 1999;76:370–380.
39 Oetting WS: Albinism. Curr Opin Pediatr 1999;11:565–571.
40 Kivitie-Kallio S, Norio R: Cohen syndrome: Essential features, natural history and heterogeneity. Am J Med Genet 2001;102:125–135.
41 Cregg M, Woodhouse JM, Pakeman VH, Saunders KJ, Gunter HL, Parker M, Fraser WI, Sastry P: Accommodation and refractive error in children with Down syndrome: Cross-sectional and longitudinal studies. Invest Ophthalmol Vis Sci 2001;42:55–63.
42 Mao JR, Bristow J: The Ehlers-Danlos syndrome: On beyond collagens. J Clin Invest 2001;107:1063–1069.
43 McKusick VA: Heritable Disorders of Connective Tissue, ed 4. St Louis, Mosby, 1972.
44 Montgomery RA, Geraghty MT, Bull E, Gelb BD, Johnson M, McIntosh I, Francomano CA, Dietz HC: Multiple molecular mechanisms underlying subdiagnostic variants of Marfan's syndrome. Am J Hum Genet 1998;63:1703–1711.
45 Allen RA, Straatsma BR, Apt l, Hall MO: Ocular manifestations of the Marfan syndrome. Trans Am Acad Ophthalmol Otolaryngol 1967;71:18–38.
46 Ayme S, Preus M: The Marshall and Stickler syndromes: Objective rejection of lumping. J Med Genet 1984;21:34–38.
47 Stickler GB, Belau PG, Farrell FJ, Jones JD, Pugh DG, Steinberg AG, Ward LE: Hereditary progressive arthro-ophthalmopathy. Mayo Clin Proc 1965;40:433–455.
48 Stickler GB, Hughes W, Houchin P: Clinical features of hereditary progressive arthro-ophthalmopathy (Stickler syndrome): A survey. Genet Med 2001;3:192–196.
49 Bear JC: Epidemiology and genetics of refractive anomalies; in Grosvenor T, Flom MC (eds): Refractive Anomalies: Research and Clinical Applications. Boston, Butterworth-Heinemann, 1991, pp 57–80.
50 Karlsson JL: Influence of the myopia gene on brain development. Clin Genet 1975;8:314–318.
51 Franceschetti A: Haute myopie avec décollement retinien héréditaire. J Hum Genet 1953;2:283–284.

52 Grice KM, DelBono EA, Haines JL, Wiggs JL, Gwiazda JE: Genetic analysis of pedigrees affected by juvenile-onset myopia. Invest Ophthalmol Vis Sci 1996;37(suppl):4603.
53 DelBono EA, Kern J, Grice K, Haines JL, Gwiazda JE; Wiggs JL: Genetic studies of juvenile-onset myopia. Invest Ophthalmol Vis Sci 1999;40(suppl):3667.
54 Wu H, Yong S, Tan E, Yap E: Further evidence of association between trabecular meshwork-induced glucocorticoid response (TIGR/Myocilin) gene and severe myopia. Invest Ophthalmol Vis Sci 2000;41(suppl):170.
55 Young TL, Guo XD, King RA, Rada JA: Identification of genes expressed in a human scleral cDNA library. Invest Ophthalmol Vis Sci 2002;43(suppl):2466.
56 Haim M, Fledelius HC, Skarsholm: X-linked myopia in a Danish family. Acta Ophthalmol 1988;66:450–456.
57 Young TL, Ronan SM, Alvear AB, Peterson JM, Dewan AT, Holleschau AM, King RA: Mapping refinement and mutation analysis of syndromic X-linked high myopia. Am J Hum Genet 2000;67(suppl):1806.
58 Young TL, Ronan SM, Drahozal LA, Wildenberg SC, Alvear AB, Oetting WS, Atwood L, Wilkin DJ, King RA: Evidence that a locus for familial high myopia maps to chromosome 18p. Am J Hum Genet 1998;63:109–119.
59 Young TL, Ronan SM, Alvear AB, Wildenberg SC, Oetting WS, Atwood LD, Wilken DJ, King RA: A second locus for familial high myopia maps to chromosome 12q. Am J Hum Genet 1998;63: 1419–1424.
60 Li SQ, Zhang QJ, Xiao XS, Chu GJ, Guo XM, Hang FS, Li JZ, Guo L, Jia XY: Mutational screening of decorin gene in 90 Chinese probands with high myopia. Invest Ophthalmol Vis Sci 2001;42 (suppl):2123.
61 Young T, Paluru P, Heon E, Bebchuck K, Armstrong C, Ronan S, Holleschau A, Petersen, J, Alvear A, Wildenberg S, King R: A new locus for autosomal dominant high myopia maps to chromosome 17q21–23. Am J Hum Genet 2001;69(suppl):2022.
62 Naiglin L, Gazagne C, Dallongeville F, Thalamas C, Ider A, Rascol O, Malecaze F, Calvas P: A genome wide scan for familial high myopia suggests a novel locus on chromosome 7q36. J Med Genet 2002;39:118–124.
63 Williams RW, Zhou G: Development of genetic models for myopia research: High-resolution mapping of a new set of loci that control eye size in mice. Invest Ophthalmol Vis Sci 2001;42(suppl):3508.

Prof. Dr. rer. nat. Frank Schaeffel
Section of Neurobiology of the Eye,
University Eye Hospital Tübingen, Dept II (Chair: Prof. Zrenner),
Calwerstrasse 7/1, D–72076 Tübingen (Germany)
Tel. +49 70 71 2980739, Fax +49 70 71205196, E-Mail frank.schaeffel@uni-tuebingen.de

Wissinger B, Kohl S, Langenbeck U (eds): Genetics in Ophthalmology.
Dev Ophthalmol. Basel, Karger, 2003, vol 37, pp 50–66

A Molecular Perspective on Corneal Dystrophies

A.L. Vincent[a], *D. Rootman*[a,b], *F.L. Munier*[c], *E. Héon*[a,b]

[a] Department of Ophthalmology, The Hospital for Sick Children,
[b] Toronto Western Hospital, Vision Science Research Program, Toronto, Canada and
[c] Unité d'oculogénétique, Hôpital Jules Gonin, Université de Lausanne, Switzerland

Abstract

Corneal dystrophies refer to a group of corneal diseases and that are genetically determined. These have been traditionally classified with respect to the layer of cornea involved. We now know that this does not reflect the underlying pathobiology. Most of the corneal dystrophies are of Mendelian inheritance with some phenotype diversity and a variable degree of penetrance. The dystrophies involving enzymatic processes tend to be of autosomal recessive inheritance. In some cases, such as keratoconus, the inheritance pattern is not always clear and is considered complex. The age of onset of the disease, as in most inherited eye disorders, is variable and does not reflect the underlying pathogenic defect. Few cases are congenital.

Our understanding of corneal dystrophies is undergoing somewhat of a revolution as over 12 chromosomes have been associated with corneal dystrophies with mutations identified in at least 14 genes if one includes anterior segment dysgenesis in this group of conditions. Several dystrophies remain without a gene or a genetic location (locus) and more familial studies are required. The new molecular information is challenging the traditional thinking about these conditions that was usually guided by the histopathological findings. As this new knowledge becomes more refined, the classification of this group of disorders will eventually be revisited to have a molecular basis. The elucidation of the underlying biochemical pathways may allow us to envisage the possibility of modulating these phenotypes in the future.

Introduction

This chapter will discuss selected primary corneal dystrophies for which molecular information is now available (table 1). We will also briefly discuss conditions such as keratoconus and Peters' anomaly, as these can be genetically

Table 1. Genes involved in corneal dystrophies

Disease	Inheritance	Corneal layer involved	Chromosome	Gene	Putative role/Function
Meesman	AD	Epithelium	12q	*K3*	Expressed in corneal epithelium Intermediate filament assembly
	AD	Epithelium	17q	*K12*	Fragility of keratinocytes Intracellular keratin aggregation
Reis-Bückler	AD	Bowman's mbr	5q	*TGFβI*	
Thiel-Behnke	AD	Bowman's mbr	5q	*TGFβI*	Keratoepithelin-related amyloid deposits
	AD	Bowman's mbr	10q23–q24	?	
Granular	AD	stroma		*TGFβI*	Keratoepithelin-related amyloid deposits
Lattice I	AD	Stroma	5q	*TGFβI*	
Lattice II	AD	Stroma	9	*Gelsolin*	Gelsolin-related amyloid deposits
Lattice IIIa	AD	Stroma	5q	*TGFβI*	Keratoepithelin-related amyloid deposits
Avellino	AD	Stroma	5q	*TGFβI*	Keratoepithelin-related amyloid deposits
Macular	AR	Stroma	16	*CHST6*	Abnormal sulfated keratan sulfate
Gelatinous drop-like	AR	Stroma	1p32	*M1S1 (TACS-TD2)*	Tumor-associated antigen, truncated protein results leads to amyloid deposition
Fuchs'	AD	Endothelium	1p34–p32	*COL8A2*	Structural role?
PPD	AD	Descemet mbr/ endothelium	20p11–q11, 1p34	*COL8A2, VSX1*	Structural role? Developmental role?
Peters' anomaly	AR	NS	2p	*CYP1B1*	Hydroxylation of 17β-estradiol
	?			*PAX6, PITX2*	Transcription factors
ASMD	AD	NS		*PITX3*	Transcription factors
Cornea plana	AD, AR	NS	12q21	*KERA*	Neural crest cell development Maintenance of transparency?

mbr = Membrane, AD = autosomal dominant, AR = autosomal recessive, ASMD = anterior segment mesodermal dysgenesis, NS = non-specific.

determined and molecular information is available. For a comprehensive clinical and historical review of corneal dystrophies, the reader is referred to classic texts [1–3].

Development, Structure and Function

The cornea forms between 5 and 6 weeks of gestation into six concentric layers (fig. 1). The outer epithelium is anchored to a basement membrane, over

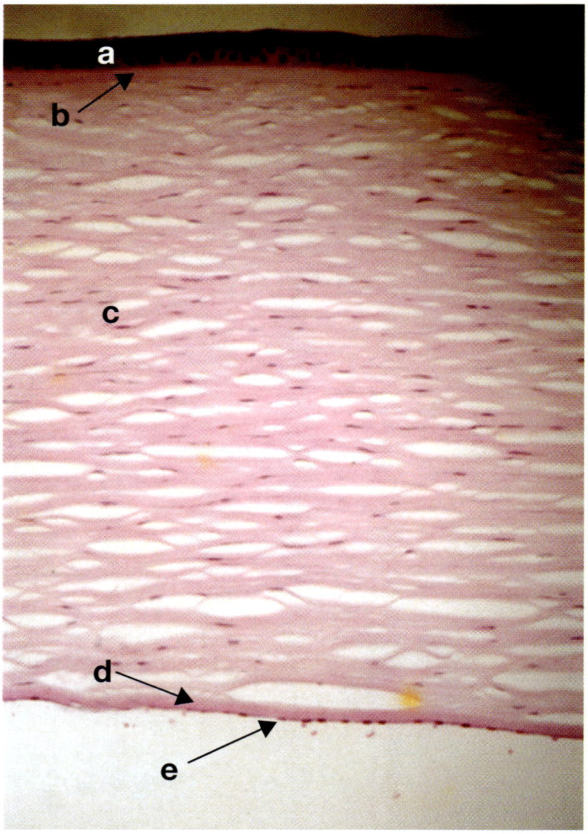

Fig. 1. Hematoxylin and eosin stain of a normal cornea (courtesy of the Eye Bank of Canada). a = Stratified epithelium, b = Bowman membrane, c = stroma, d = Descemet's membrane, e = corneal endothelium.

the acellular Bowman's layer anterior to the stroma. Constituents of this layer are believed to be both synthesized and secreted by epithelial cells and stromal keratocytes. The posterior corneal stroma is lined by the collagenous Descemet's membrane which is lined by a monolayer of endothelial cells which permits the passage of nutrients from the aqueous humor into the cornea and is responsible for maintaining the relatively low level of stromal hydration necessary for corneal transparency. Stromal hydration is controlled by the activity of ionic pumps in the plasma membrane of endothelial cells.

The adult cornea has an average diameter of 12.6 mm horizontally and 11.7 mm vertically. Centrally, the thickness measures 0.52 mm, increasing towards the periphery. The major types of cytoplasmic filaments include keratin (intermediate filaments), actin and microtubules, keratin being predominant. The intermediate filaments are formed by pairing of two specific keratin proteins; K5 and K14 in the basal cells, and K3 and K12 in the suprabasal cells [4]. The stroma represents 90% of the corneal thickness, and consists of highly uniform collagen fibrils (22.5–32 nm diameter), predominantly type I, III, V, VI, XII and XIV cross-linking fibrillar collagen forming microfibril networks [4] with the keratocytes in between. The keratocytes secrete the extracellular matrix around the collagen consisting of acidic, negatively charged proteoglycans, with keratan sulfate and dermatan sulfate predominating [4, 5]. The proteoglycans play a role in maintaining the regular collagen fibril spacing. The major functions of the stroma are to maintain the proper curvature of the cornea, to provide mechanical resistance to intraocular pressure, and to transmit light into the eye without significant absorbance. Adult Descemet's membrane contains fibronectin, laminin, type IV and type VIII collagen, heparan sulfate, and dermatan sulfate proteoglycan [4].

The Dystrophies (table 1)

Meesman's Corneal Dystrophy (OMIM 122100)

The inheritance is autosomal dominant with an onset in early childhood (~12 months). A myriad of intraepithelial vesicles/microcysts increase in number throughout life which can be associated with recurrent punctate erosions. Symptoms are variable, ranging from asymptomatic to contact lens intolerance, pain and lacrimation associated with erosions, intermittent blurred vision and photophobia. Vision impairment is usually mild and superficial keratectomy is rarely required.

On histopathology, the epithelium is irregularly thickened with numerous cytoplasmic vacuolations and intraepithelial cysts containing PAS-positive 'peculiar substance' stained with Alcian blue and colloidal iron stains and suggestive of keratin on electron microscopy (EM).

The original Meesman's German pedigree was mapped to chromosome 12q12–q13 [6], whereas a family from Northern Ireland was mapped to chromosome 17q12 [6]. Mutations have been identified in both *K12* (OMIM 601687) on chromosome 12q12–q13 and *K3* (OMIM 148043) on chromosome 17q12. Both genes are expressed in the anterior corneal epithelium. K3 and K12 contain a highly conserved helix boundary motif, which plays a critical role for structural integrity and filament assembly [6–8]. The mutations may have a dominant negative effect and lead to aberrant assembly of intermediate filaments with resultant disruption of keratinocyte filament architecture, and cell lysis following mild trauma [7]. *K12* knock-out mice

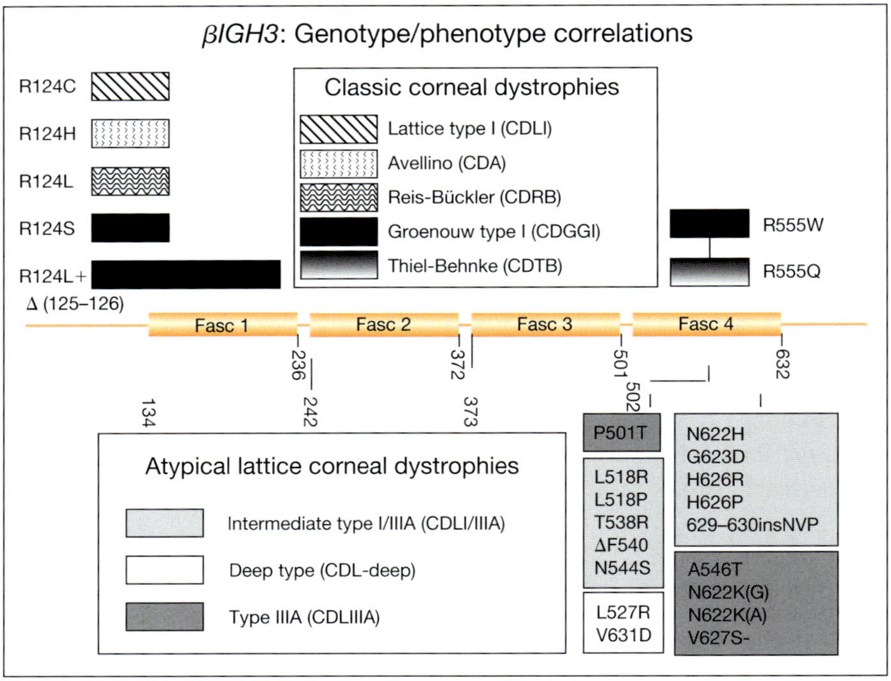

Fig. 2. *βIGH3* (TGFβI) phenotype-genotype correlations. Schematic representation of the *βIGH3* (TGFβI) gene with the four fasc domains and the respective grouping of mutations.

show complete loss of the K3/K12 cytoskeleton with resultant fragility of keratinocytes [6].

Lisch Corneal Dystrophy

Inheritance is X-linked dominant with an onset by the second decade. The band-shaped gray opacities and densely crowded large whorled microcysts of the corneal epithelium are distinct genetically and histopathologically from Meesman's corneal dystrophy [9–11]. A progressive decrease in vision associated with opacification but no pain has been reported with corneal erosions. Improvement can be seen with soft contact lens wear [11] and/or corneal scrapping.

There was suggestion of linkage to Xp22.3 and the K3 and K12 loci were genetically excluded [11].

Reis-Bücklers ('Geographic') Corneal Dystrophy, CDBI (OMIM 121900)

Inheritance is autosomal dominant with complete penetrance but variable severity. This condition usually manifests in the first decade of life with variable forms of reticular gray-white opacification at the level of Bowman's layer giving a ground-glass appearance in intervening areas. The entire cornea is involved, but most densely axially. Irregularities of the corneal surface can lead to recurrent corneal erosions with reduced corneal sensation. Recurrent attacks of photophobia and irritation become less frequent with increasing age. Progressive visual loss is due to corneal opacification. This can be managed with debridement, superficial keratectomy or phototherapeutic keratectomy [12]. When severe, lamellar or penetrating keratoplasty can be indicated but recurrence in graft can be seen early. The opacified Bowman's layer is replaced by multistratified PAS-positive (Masson's) eosinophillic material, with projections into epithelium and in the anterior stroma [13] but do not involve the epithelial basement membrane. EM shows tubular microfibrils, crescent or rod-shaped bodies, interspersed between collagen in Bowman's layer. Mutations were identified in transforming growth factor, β-induced (*TGFβI* or *βIGH3*) (OMIM 601692) on chromosome 5q31 (fig. 2). Despite a strong allelic heterogeneity, some strong phenotype-genotype correlations are observed such as the R124L change which is specific to the Reis-Bückler phenotype.

Different mutations in *TGFβI* may also cause granular corneal dystrophy type I (GCD1), GCDII and III, lattice corneal dystrophy type I (LCDI), LCDIIIa, intermediate type LCDI/LCDIII and LCD-deep, as well as Thiel-Behnke dystrophy [14] (see below). Mutations involve CpG dinucleotides. *TGFβI* is expressed in keratocytes and encodes for keratoepithelin, a highly conserved 683-amino-acid protein. This protein contains an N-terminal secretory signal, 4 domains of internal homology, and an arg-gly-asp (RGD) motif at the C terminus, which is found in many extracellular matrix proteins. The RGD motif modulates cell adhesion, and acts as a recognition sequence for integrin binding. Mutations in gene result in progressive accumulation of corneal deposits shown to contain keratoepithelin. Aggregation of abnormal isoforms of keratoepithelin are associated with amyloid or other non-fibrillar deposits depending on the location and nature of the mutation.

Honeycomb-Shaped Dystrophy/Thiel-Behnke Dystrophy (CBDII) (OMIM 602082)

This autosomal dominant dystrophy usually manifests during the second decade with subepithelial axial honeycomb-like opacities with clear corneal periphery. Recurrent corneal erosions can be manifest until the fourth to fifth decades. The secondary progressive visual loss can reach 20/100. However, the corneal surface is smooth and the corneal sensation normal. The epithelial basement membrane and Bowman's layer may be focally absent. 'Curly' collagen fibers seen on EM [13] correspond to irregular epithelial and subepithelial PAS-positive fibrocellular deposits. Superficial keratectomy, lamellar or penetrating keratoplasty can be indicated. Recurrence in the graft may occur but later than in CBDI.

This phenotype is genetically heterogeneous with mutations in *TGFβI* R555Q (fig. 2), and some families mapped to chromosome10q24 [15] for which the gene is not yet identified.

Stromal

Granular Dystrophy (Groenouw Type 1) (OMIM 121900)

This autosomal dominant dystrophy shows complete penetrance and variable expressivity. The onset of signs occurs in the first and second decade with discrete white granular opacities in the central cornea, within anterior stroma, that may resemble breadcrumbs with a clear intervening stroma. With time, the opacities increase in number, density, size and depth. The peripheral cornea remains clear whereas the intervening cornea becomes like ground glass. Surface irregularities may develop and sometimes lead to corneal erosions and intense pain. Vision progressively decreases as a result of the scarring and increase in density of the deposits usually by the fourth and fifth decade. Lamellar or penetrating keratoplasty may be required but recurrences may occur early.

Deposits described as 'hyaline', stain bright red with Masson's trichrome. On EM, rod and trapezoid deposits extend into more posterior layers. Mutations in *TGFβI* show strong phenotype-genotype correlations. R555W is a definite hot spot but R124S is also seen in these patients [14].

Lattice Corneal Dystrophy

Various subtypes of lattice corneal dystrophy are distinguished by the variability of phenotype severity or the association of systemic findings. These distinctions are genetically determined [14] (fig. 2).

Lattice Corneal Dystrophy Type 1 CDL1 (OMIM 122200)

CDL1 has an autosomal dominant inheritance with complete penetrance and phenotypic variability. The onset is usually during the first decade with anterior subepithelial white dots and refractile filamentous linear amyloid deposits +/− nodules within stroma. Later, line branching becomes thicker with radial orientation and involves deeper stoma. Progressive opacification involves the central visual axis with clouding of intervening stroma. Frequent recurrent erosions may present at a young age and can be helped with topical lubrication, patching, therapeutic contact lenses. When there is scarring, phototherapeutic keratectomy and penetrating keratoplasty may improve vision but recurrences in graft are common.

On light microscopy, the epithelium is irregular with a thickened basement membrane, and a fragmented Bowman's layer. Fibrillar deposits in anterior stromal layers extend posteriorly, stain intensely with Congo red and show birefringence and dichroism. The R124C TGFβ1 mutation is disease specific [14] (fig. 2).

Lattice Corneal Dystrophy Type II (Familial Amyloidosis, Meretoja Syndrome, Finnish Type, OMIM 105120)

This rare autosomal dominant dystrophy manifests in early adulthood with also systemic findings unlike the other forms of lattice dystrophy. The ocular signs include lattice lines – fewer than LCDI, more radial orientation peripherally with relative central sparing. Reduced corneal sensitivity and recurrent epithelial erosions manifest after the age of 40 years with possible scarring and eventual reduced visual acuity. Dry eyes, pain and lacrimation are associated with erosions. Overall, symptoms are less severe than LCDI. Systemic findings include lax skin, peripheral neuropathy and cardiomyopathy.

The formation of subepithelial scar tissue is associated linear amyloid material under Bowman's layer, in anterior and midstroma [16]. The amyloid fibrils correspond to an internal degradation product of gelsolin leading to a progressive loss of corneal sensory nerves with decrease in sensation of cornea, skin and cranial nerves predominantly. LCDII maps to chromosome 9q34 and residue D187 of the gelsolin gene (*GSN*) (OMIM 137350) represents a hot spot for disease-causing mutations (D87N, D187Y). *GSN* is widely expressed and encodes for an actin filament modulating (severing and capping) protein [17] which exerts its action in the presence of submicromolar calcium. The amyloid protein in the Finnish type is a fragment of the actin-filament-binding region of a variant gelsolin molecule [18, 19].

Avellino Corneal Dystrophy (CDA)

CDA is autosomal dominant with complete penetrance and shows highly variable expressivity. It manifests in the second decade with both granular and amyloid linear branching deposits within stroma. Granular opacities have an earlier onset than the amyloid deposits, and are located more superficially. Progressive opacification of the central visual axis by deposits may decrease vision enough to require phototherapeutic keratectomy or penetrating keratoplasty. The granular subepithelial to midstromal deposits stain with Masson trichrome. The fibrillar or fusiform, deep stromal deposits containing amyloid stain with Congo red (birefringent) [20, 21]. The R124H mutation of *TGFβI* is specific to this condition [14] (fig. 2).

Macular Dystrophy (MCDI (OMIM 217800), MCDIa, MCDII)

The different subtypes of macular dystrophy are genetically and biochemically determined. The inheritance is autosomal recessive inheritance with onset in the first decade of life and no significant variability in the phenotype. Early, the fine opacities have indistinct edges, starting axially in superficial stroma. The intervening stroma has a ground-glass appearance. Later, opacities extend peripherally and into deep stroma. The corneal surface becomes irregular with decreased corneal sensation and eventual corneal thinning. Irritation and progressive loss of vision can become severe by the third decade and may require corneal graft with good results.

The characteristic accumulation of glycosaminoglycans (GAGs) stains with Alcian blue and colloidal iron. MCDI is characterized by the absence of keratan chain sulfation (KCS) in cornea and cartilage and no appreciable serum or corneal KCS. In MCDII, serum and corneal keratan sulfate are detectable, may be reduced but are often normal. MCD was mapped to chromosome 16q22 and disease-causing mutations involve *CHST6* (OMIM 603797). *CHST6* encodes an enzyme; carbohydrate sulfotransferase [22], which is expressed in the cornea, also trachea and spine. The gene product, corneal N-acetylglucosamine-6-sulfotransferase (C-GlcNAc6ST), initiates sulfation of keratan sulfate in cornea [23]. MCDI is due to mutations (missense, deletion, insertions, frameshift) in coding regions of *CHST6* [22, 24]. These result in synthesis of an inactive enzyme with the synthesis and secretion of proteoglycans substituted with polylactosamine instead of keratan sulfate. Carrier state is high in Iceland [24]. MCDIa was seen in families from Saudi Arabia where there is absence of keratan sulfate in corneal stroma and serum, but presence within the keratocyte. The genetic changes in MCDII are not intragenic but involve *CHST6* deletions/rearrangements of upstream regions thought to contain gene regulatory elements. These changes affect *CHST6* transcription [22] and reduced sulfation resulting in premature keratan sulfate chain termination [23].

MCDI and II both have accumulation of other GAGs (chondroitin/dermatan sulfate/hyaluron). Therefore, the corneal opacity may result from not only lack of KCS, but deposition of extra GAGs which may interfere with collagen fibril arrangement [23]. MCDI and II can occur in the same family [22, 25]. The MCDII genotype is dominant over MCDI if a compound heterozygote has a coding mutation and upstream mutation.

Gelatinous Drop-Like Corneal Dystrophy (Amyloid) (GDLD; OMIM 204870)

GDLD is also autosomal recessive with an onset of signs in the first decade with flattish subepithelial nodular deposits similar to early band keratopathy.

Later, there is an increase in number and depth of nodular deposits to become raised yellow-gray gelatinous masses (mulberry), with surrounding dense subepithelial opacities. The severe corneal amyloidosis can lead to blindness. Recurrent lamellar keratoplasty or penetrating keratoplasty may be required. The incidence in Japan is of 1 in 300,000.

The subepithelial amyloid deposits also involve the anterior stroma, and may be found at a depth of two-thirds of the cornea and in conjunctival stroma [20].

Genetic linkage studies mapped the disease to chromosome 1p31. Mutations were identified in *M1S1* (formerly *TROP2* and *GA733-1*), encoding a gastrointestinal tumor-associated antigen. TACSTD2 (OMIM 137290), tumor-associated calcium signal transducer 2 [26], transduces an intracellular calcium signal, and acts as a cell surface receptor [17]. In GDLD, the corneas have high epithelial permeability, which directly correlates with abnormalities in epithelial structure, including irregular cell junctions. This suggests that the abnormal *M1S1* gene product may affect epithelial cell junctions resulting in increased cell permeability in GDLD corneas [27]. Q118X is a recurrent mutation, present in 82% of the Japanese GDLD population [17, 28]. In 2 unrelated patients from India, a Q118E mutation was identified potentially indicating Q118 as a mutational hotspot. Locus heterogeneity is reported after exclusion of the M1S1 region by linkage analysis [29].

Central Crystalline Dystrophy (of Schnyder) (OMIM 121800)

This autosomal dominant dystrophy has an early onset in the first decade with central oval or annular corneal clouding with an irregular edge and a clear periphery. A prominent arcus lipides is seen with multiple fine needle-like iridescent crystal just posterior to Bowman's layer. This may resemble the corneal dystrophy of cystinosis. The progression is minimal with crystals extending into deeper stromal layers [30] with slowly progressive vision loss. Phototherapeutic keratectomy can be beneficial if deposits are superficial [12].

The birefringent crystals are composed of phospholipid (sphingomyelin), unesterified cholesterol and cholesterol esters [31] stain with Oil Red O (frozen specimen). Bowman's layer is absent except in periphery [32]. Skin biopsy demonstrates abnormal membrane-bound vacuoles in fibroblasts which suggests abnormal cholesterol metabolism. Linkage analysis of two kindreds of Swede-Finn descent from central Massachusetts identified a 16-cM disease gene interval at 1p36–p34.1 [33].

Bietti Crystallin Corneoretinal Dystrophy (OMIM 210370)

This autosomal recessive dystrophy manifests in the third decade with crystals involving the cornea, retina and lymphocytes. The corneal crystals are paralimbal and subepithelial or anterior stromal. The retinal crystals are diffusely scattered and are associated with a progressive retinal degeneration of photoreceptors and RPE. Most of the visual impairment comes from the retinal disease, leading to legal blindness in the fifth to sixth decade.

On light microscopy, crystals are seen within fibroblasts, or extracellular matrix [34], adjacent to complex lipid inclusions in keratocytes and conjunctival fibroblasts. On EM, crystals resemble cholesterol or cholesterol esters. In a study of 10 families with BCD, Jiao et al. [35] reported linkage to chromosome 4q35-qter (maximum lod of 5.3, θ = zero).

Endothelial/Descemet's Membrane

Posterior Polymorphous Dystrophy (PPD)
(OMIM 122000, 120252, 605020)

This autosomal dominant dystrophy shows variable expression and variable age of onset. Although it is usually a disease of adulthood, PPD can be severe and present at birth. Changes consist of a variable degree of vesicular endothelial lesions and/or with basement membrane thickening. This may be localized or more diffuse and associated with corneal edema. Vision loss is usually not significant but is highly variable. Corneal edema can develop to a degree necessitating a corneal graft. There is also an increased risk for glaucoma and keratoconus [36].

The abnormal anterior banded layer of Descemet's membrane is lined posteriorly by an abnormal posterior collagenous layer. Multilayering of endothelial cells is seen in periphery [37] with metaplasia and epithelialization of endothelial cells [38]. There is a mosaic of better-preserved and dystrophic multilayered endothelial cells in the presence or absence of the normal components of Descemet's membrane.

PPD is genetically heterogeneous with mutations identified in *VSX-1* (chromosome 20p11.2–20q11.2) [39, 40] and *COL8A2* (chromosome 1p34.3–p32) [41]. Data suggest that the chromosome 20-related PPD is an allelic variant of keratoconus. *VSX1* appears to play a role in ~9% of PPD cases and 4.5% of keratoconus cases. The chromosome 1-related PPD is an allelic variant of Fuchs' endothelial dystrophy. *Col8A2* could play a role in ~6% of cases of PPD.

Fuchs' Endothelial Dystrophy (OMIM 136800)

This is the most common primary disorder of corneal endothelium which can be sporadic or autosomal dominant [42] with variable expression. The onset of signs and symptoms is usually from the fourth decade onwards with central cornea guttata, little wart-like excrescences of Descemet's membrane, beaten metal appearance progressing to stromal folds, corneal edema and endothelial polymegathism. Later, the visual loss can be significant and painful because of corneal decompensation. Penetrating keratoplasty of cases of Fuchs' endothelial dystrophy accounts for up to 19% of corneal grafts with a success rate in these patients of over 90%.

Some families have been mapped to chromosome 1p34.3–p32 for which mutations have been identified in *COL8A2* (3.4%) [41]. The α2 subunit of type VIII collagen is a member of a family of extracellular matrix proteins [43] and contains an important triple helix repeat, a proline-rich region which is a site of hydroxylation. Mutations within triple helical domain theoretically disrupts stability of supramolecular assembly [41] and have been associated with PPD (~6%) and FECD (~3.4%). The phenotype-genotype correlation is poor as the same mutation can lead to different phenotypes. Mitochondrial mutations have been documented in a case also affected with sensorineural hearing loss, diabetes, cardiac conduction defects, ataxia and hyperreflexia [44].

CHED (Congenital Hereditary Endothelial Dystrophy) – CHED1 (OMIM 121700), CHED2 (OMIM 217700)

CHED1 is of autosomal dominant inheritance with onset at birth or the first few months of life, up to 8 years [45]. CHED2 is of autosomal recessive inheritance with an early onset of signs and symptoms at birth or within the first few weeks of life. The cornea has a ground-glass appearance and the corneal epithelium may be roughened. There is no guttatae and the corneal sensitivity is normal corneal. Although the decrease in vision is moderate to severe, nystagmus is uncommon.

In CHED1, the stromal edema is homogeneous with or without spheroidal (droplet) degeneration. Descemet's membrane is thickened, but not necessarily reduplicated [45]. The endothelium is vacuolated and focally absent. In CHED2, spheroidal degeneration is not common and the basement membrane is more thickened than in CHED1.

CHED is genetically heterogeneous with CHED1 linked to 2.7 cM region 20p11.2–20q11.2 [46] and CHED2 linked to 20p13 [47]. Other CHED loci remain to be identified.

Developmental and Other 'Corneal Dystrophies'

Corneal opacification, central (leucoma) or peripheral (sclerocornea), may manifest in various forms of dysgenesis of the anterior segment. An example is Peters' anomaly for which mutations are identified in the eye development genes such as *PAX6* [48], *PITX2* [49], *FOXC1* [50] and *CYP1B1* [51].

Mutations in *PAX6* can also produce autosomal dominant keratitis [52]. Sclerocornea has been described occurring in autosomal dominant and recessive pedigrees [53]. Cornea plana (CNA1 (OMIM 121400), CNA2 (OMIM 217300)) has two subtypes distinguished by their inheritance pattern and severity. CNA1 is autosomal dominant whereas CNA2 is autosomal recessive and more severe. Both are present at birth and are related to an abnormal curvature of the cornea. CNA1 and CNA2 were linked to chromosome 12q within 3 cM of keratocan *(KERA)* located at 12q22, [54]. *KERA* mutations were found in patients with CNA2 and one family affected with CNA1. No mutations were found in one of the original CNA1 families [55]. There is evidence for involvement of a second locus [54, 55]. KERA is a keratan sulfate proteoglycan, a member of the small leucine-rich proteoglycan family (SLRP). These are highly evolutionarily conserved with other SLRPs including lumican and mimecan and are important for development and maintenance of corneal transparency and structure [55, 56]. *KERA* has a very restricted expression in early neural crest development, and later in corneal stromal cells [56].

Keratoconus (OMIM 148300)

Keratoconus can be sporadic or autosomal dominant in 6–8% [57]. Its prevalence in first-degree relatives is 15–67 times higher than the general population [58] and it has been observed in identical twins [59]. The onset is around puberty with a progressive ectatic dystrophy leading to corneal thinning, with induced irregular myopic astigmatism which may be markedly asymmetrical [60]. Corneal topography is useful for diagnosis showing an increase in the central K's. In advanced cases, anterior scarring can be seen and hydrops may occur when Descemet's membrane ruptures with subsequent epithelial and stromal edema. The decreased vision associated with hydrops or corneal scarring may require a corneal graft. For comprehensive review, see Rabinowitz [60]. Keratoconus is a genetically determined disorder with a combination of biochemical, structural and cellular changes [61–64]. It has been associated with several chromosomal anomalies; trisomy 21, Turner's syndrome, ring chromosome 13, translocation 7;11, connective tissue disorders; Ehlers-Danlos, Marfan syndrome, osteogenesis imperfecta, mitral valve prolapse and other ocular diseases such as Leber's congenital amaurosis, and atopy [60]. Mutations in the *VSX-1* transcription factor were identified in 4.7% of patients with isolated keratoconus [40]. This gene also plays a role in posterior polymorphous dystrophy (see above).

Systemic Abnormalities with Corneal Manifestations

Inherited systemic diseases associated with significant corneal changes include X-linked ichthyosis [65], lecithin cholesterol acyltransferase deficiency

[66], mucopolysaccharidosis, and Fabry's disease among others. These discussions are beyond the scope of this chapter.

Conclusions

Molecular ophthalmology is redefining our understanding of inherited corneal disorders. This new knowledge outlines that although conditions may be clinically distinct, they may share a common genetic background. Various categories of genes are being identified to play a role in the determination of corneal transparency and genetically-determined corneal diseases. These include those involved in the regulation of eye and corneal development, as well as factors that determine and maintain ultrastructural corneal arrangement and its metabolic homeostasis. These genes are part of molecular pathways being characterized, some of which are likely play a role beyond the cornea. The understanding of these pathways will be critical to the potential modulation of phenotypes. The genetic studies of corneal dystrophies are evolving and are being an efficient approach in tying these pathways together and defining new therapeutic opportunities.

References

1 Chang SW, Tuli S, Azar DT (eds): Corneal Dystrophies, ed 1. New York, Oxford University Press, 1998.

2 Bajart AM, Pavilack MA: Corneal manifestations of metabolic disease; in Albert DM, Jakobiec FA (eds): Principles and Practice of Ophthalmology, ed 2. Philadelphia, Saunders, 2000, vol 2, pp 756–781.

3 Starck T, Hersch PS, Kenyon KR: Corneal dysgeneses, dystrophies and degenerations; in Albert DM, Jakobiec FA (eds): Principles and Practice of Ophthalmology, ed 2. Philadelphia, Saunders, 2000, vol 2, pp 695–755.

4 Panjawi N: Cornea and sclera; in Harding J (ed): Biochemistry of the Eye. London, Chapman & Hall, 1997, pp 16–51.

5 Pepose J, Ubels J: The cornea; in Harts W Jr (ed): Adler's Physiology of the Eye, ed 9. St Louis, Mosby-Year Book, 1992, pp 29–70.

6 Irvine AD, Corden LD, Swensson O, Swensson B, Moore JE, Frazer DG, Smith FJ, Knowlton RG, Christophers E, Rochels R, Uitto J, McLean WH: Mutations in cornea-specific keratin K3 or K12 genes cause Meesman's corneal dystrophy. Nat Genet 1997;16:184–187.

7 Coleman CM, Hannush S, Covello SP, Smith FJ, Uitto J, McLean WH: A novel mutation in the helix termination motif of keratin K12 in a US family with Meesman corneal dystrophy. Am J Ophthalmol 1999;128:687–691.

8 Corden LD, Swensson O, Swensson B, Rochels R, Wannke B, Thiel HJ, McLean WH: A novel keratin 12 mutation in a German kindred with Meesman's corneal dystrophy. Br J Ophthalmol 2000;84:527–530.

9 Lisch W, Steuhl KP, Lisch C, Weidle EG, Emmig CT, Cohen KL, Perry HD: A new, band-shaped and whorled microcystic dystrophy of the corneal epithelium. Am J Ophthalmol 1992;114:35–44.

10 Robin SB, Epstein RJ, Kornmehl EW: Band-shaped, whorled microcystic corneal dystrophy. Am J Ophthalmol 1994;117:543–544.

11 Lisch W, Buttner A, Oeffner F, Boddeker I, Engel H, Lisch C, Ziegler A, Grzeschik K: Lisch corneal dystrophy is genetically distinct from Meesman corneal dystrophy and maps to xp22.3. Am J Ophthalmol 2000;130:461–468.
12 Dinh R, Rapuano CJ, Cohen EJ, Laibson PR: Recurrence of corneal dystrophy after excimer laser phototherapeutic keratectomy. Ophthalmology 1999;106:1490–1497.
13 Kuchle M, Green WR, Volcker HE, Barraquer J: Re-evaluation of corneal dystrophies of Bowman's layer and the anterior stroma (Reis-Bucklers and Thiel-Behnke types): A light and electron microscopic study of eight corneas and a review of the literature. Cornea 1995;14:333–354.
14 Munier FL, Frueh BE, Othenin-Girard P, Uffer S, Cousin P, Wang MX, Héon E, Black GCM, Blasi MA, Balestrazzi E, Lorenz B, Escoto R, Barraquer R, Hoeltzenbein M, Gloor B, Fossarello M, Singh AD, Arsenijevic Y, Zografos L, Schorderet DF: BIGH3 mutations in type IIIA and intermediate type I/IIIA of lattice corneal dystrophies are Fas4-specific. Invest Ophthalmol Vis Sci 2002;43:949–954.
15 Yee RW, Sullivan LS, Lai HT, Stock EL, Lu Y, Khan MN, Blanton SH, Daiger SP: Linkage mapping of Thiel-Behnke corneal dystrophy (CDB2) to chromosome 10q23–q24. Genomics 1997;46:152–154.
16 Rosenberg ME, Tervo TM, Gallar J, Acosta MC, Muller LJ, Moilanen JA, Tarkkanen AH, Vesaluoma MH: Corneal morphology and sensitivity in lattice dystrophy type II (familial amyloidosis, Finnish type). Invest Ophthalmol Vis Sci 2001;42:634–641.
17 Tsujikawa M, Kurahashi H, Tanaka T, Nishida K, Shimomura Y, Tano Y, Nakamura Y: Identification of the gene responsible for gelatinous drop-like corneal dystrophy. Nat Genet 1999;21:420–423.
18 Maury CP, Kere J, Tolvanen R, de la Chapelle A: Finnish hereditary amyloidosis is caused by a single nucleotide substitution in the gelsolin gene. FEBS Lett 1990;276:75–77.
19 Maury CP, Baumann M: Isolation and characterization of cardiac amyloid in familial amyloid polyneuropathy type IV (Finnish): Relation of the amyloid protein to variant gelsolin. Biochim Biophys Acta 1990;1096:84–86.
20 Santo RM, Yamaguchi T, Kanai A, Okisaka S, Nakajima A: Clinical and histopathologic features of corneal dystrophies in Japan. Ophthalmology 1995;102:557–567.
21 Konishi M, Yamada M, Nakamura Y, Mashima Y: Immunohistology of kerato-epithelin in corneal stromal dystrophies associated with R124 mutations of the BIGH3 gene. Curr Eye Res 2000;21:891–896.
22 Akama TO, Nishida K, Nakayama J, Watanabe H, Ozaki K, Nakamura T, Dota A, Kawasaki S, Inoue Y, Maeda N, Yamamoto S, Fujiwara T, Thonar EJ, Shimomura Y, Kinoshita S, Tanigami A, Fukuda MN: Macular corneal dystrophy type I and type II are caused by distinct mutations in a new sulphotransferase gene. Nat Genet 2000;26:237–241.
23 Plaas AH, West LA, Thonar EJ, Karcioglu ZA, Smith CJ, Klintworth GK, Hascall VC: Altered fine structures of corneal and skeletal keratan sulfate and chondroitin/dermatan sulfate in macular corneal dystrophy. J Biol Chem 2001;276:39788–39796.
24 Liu NP, Dew-Knight S, Rayner M, Jonasson F, Akama TO, Fukuda MN, Bao W, Gilbert JR, Vance JM, Klintworth GK: Mutations in corneal carbohydrate sulfotransferase-6 gene (CHST6) cause macular corneal dystrophy in Iceland. Mol Vis 2000;6:261–264.
25 Liu NP, Baldwin J, Lennon F, Stajich JM, Thonar EJ, Pericak-Vance MA, Klintworth GK, Vance JM: Coexistence of macular corneal dystrophy types I and II in a single sibship. Br J Ophthalmol 1998;82:241–244.
26 Calabrese G, Crescenzi C, Morizio E, Palka G, Guerra E, Alberti S: Assignment of TACSTD1 (alias TROP1, M4S1) to human chromosome 2p21 and refinement of mapping of TACSTD2 (alias TROP2, M1S1) to human chromosome 1p32 by in situ hybridization. Cytogenet Cell Genet 2001;92:164–165.
27 Kinoshita S, Nishida K, Dota A, Inatomi T, Koizumi N, Elliott A, Lewis D, Quantock A, Fullwood N: Epithelial barrier function and ultrastructure of gelatinous drop-like corneal dystrophy. Cornea 2000;19:551–555.
28 Tasa G, Kals J, Muru K, Juronen E, Piirsoo A, Veromann S, Janes S, Mikelsaar AV, Lang A: A novel mutation in the M1S1 gene responsible for gelatinous drop-like corneal dystrophy. Invest Ophthalmol Vis Sci 2001;42:2762–2764.

29 Ren Z, Lin PY, Klintworth GK, Iwata F, Munier FL, Schorderet DF, El Matri L, Basti S, Reddy M, Kaiser-Kupfer MI, Hejtmancik JF: Allelic and locus heterogeneity in autosomal recessive gelatinous drop-like corneal dystrophy. Hum Genet 2002;110:568–577.
30 Bron AJ, Williams HP, Carruthers ME: Hereditary crystalline stromal dystrophy of Schnyder I: Clinical features of a family with hyperlipoproteinaemia. Br J Ophthalmol 1972;56:383–399.
31 Yamada M, Mochizuki H, Kamata Y, Nakamura Y, Mashima Y: Quantitative analysis of lipid deposits from Schnyder's corneal dystrophy. Br J Ophthalmol 1998;82:444–447.
32 Garner A, Tripathi RC: Hereditary crystalline stromal dystrophy of Schnyder. II. Histopathology and ultrastructure. Br J Ophthalmol 1972;56:400–408.
33 Shearman AM, Hudson TJ, Andresen JM, Wu X, Sohn RL, Haluska F, Housman DE, Weiss JS: The gene for Schnyder's crystalline corneal dystrophy maps to human chromosome 1p34.1–p36. Hum Mol Genet 1996;5:1667–1672.
34 Wilson DJ, Weleber RG, Klein ML, Welch RB, Green WR: Bietti's crystalline dystrophy. A clinicopathologic correlative study. Arch Ophthalmol 1989;107:213–221.
35 Jiao X, Munier FL, Iwata F, Hayakawa M, Kanai A, Lee J, Schorderet DF, Chen MS, Kaiser-Kupfer M, Hejtmancik JF: Genetic linkage of Bietti crystallin corneoretinal dystrophy to chromosome 4q35. Am J Hum Genet 2000;67:1309–1313.
36 Cibis GW, Krachmer JA, Phelps CD, Weingeist TA: The clinical spectrum of posterior polymorphous dystrophy. Arch Ophthalmol 1977;95:1529–1537.
37 Sekundo W, Lee WR, Kirkness CM, Aitken DA, Fleck B: An ultrastructural investigation of an early manifestation of the posterior polymorphous dystrophy of the cornea. Ophthalmology 1994;101:1422–1431.
38 McCartney AC, Kirkness CM: Comparison between posterior polymorphous dystrophy and congenital hereditary endothelial dystrophy of the cornea. Eye 1988;2:63–70.
39 Héon E, Mathers WD, Alward WL, Weisenthal RW, Sunden SL, Fishbaugh JA, Taylor CM, Krachmer JH, Sheffield VC, Stone EM: Linkage of posterior polymorphous corneal dystrophy to 20q11. Hum Mol Genet 1995;4:485–488.
40 Héon E, Greenberg A, Kopp KK, Rootman D, Vincent AL, Billingsley G, Priston M, Dorval KM, Chow RL, McInnes RR, Heathcote G, Westall C, Sutphin JE, Semina E, Bremner R, Stone EM: VSX1: A gene for posterior polymorphous dystrophy and keratoconus. Hum Mol Genet 2002:11:1029–1036.
41 Biswas S, Munier FL, Yardley J, Hart-Holden N, Perveen R, Cousin P, Sutphin JE, Noble B, Batterbury M, Kielty C, Hackett A, Bonshek R, Ridgway A, McLeod D, Sheffield VC, Stone EM, Schorderet DF, Black GC: Missense mutations in COL8A2, the gene encoding the α2 chain of type VIII collagen, cause two forms of corneal endothelial dystrophy. Hum Mol Genet 2001;10:2415–2423.
42 Cross HE, Maumenee AE, Cantolino SJ: Inheritance of Fuchs' endothelial dystrophy. Arch Ophthalmol 1971;85:268–272.
43 Muragaki Y, Jacenko O, Apte S, Mattei MG, Ninomiya Y, Olsen BR: The α2(VIII) collagen gene. A novel member of the short chain collagen family located on the human chromosome 1. J Biol Chem 1991;266:7721–7727.
44 Albin RL: Fuchs' corneal dystrophy in a patient with mitochondrial DNA mutations. J Med Genet 1998;35:258–259.
45 Kirkness CM, McCartney A, Rice NS, Garner A, Steele AD: Congenital hereditary corneal oedema of Maumenee: Its clinical features, management and pathology. Br J Ophthalmol 1987;71:130–144.
46 Toma NM, Ebenezer ND, Inglehearn CF, Plant C, Ficker LA, Bhattacharya SS: Linkage of congenital hereditary endothelial dystrophy to chromosome 20. Hum Mol Genet 1995;4:2395–2398.
47 Hand CK, Harmon DL, Kennedy SM, FitzSimon JS, Collum LM, Parfrey NA: Localization of the gene for autosomal recessive congenital hereditary endothelial dystrophy (CHED2) to chromosome 20 by homozygosity mapping. Genomics 1999;61:1–4.
48 Hanson IM, Fletcher JM, Jordan T, Brown A, Taylor D, Adams RJ, Punnett HH, van Heyningen V: Mutations at the PAX6 locus are found in heterogeneous anterior segment malformations including Peters' anomaly. Nat Genet 1994;6:168–173.

49 Doward W, Perveen R, Lloyd IC, Ridgway AE, Wilson L, Black GC: A mutation in the RIEG1 gene associated with Peters' anomaly. J Med Genet 1999;36:152–155.
50 Nishimura DY, Searby CC, Alward WL, Walton D, Craig JE, Mackey DA, Kawase K, Kanis AB, Patil SR, Stone EM, Sheffield VC: A spectrum of FOXC1 mutations suggests gene dosage as a mechanism for developmental defects of the anterior chamber of the eye. Am J Hum Genet 2001;68:364–372.
51 Vincent A, Billingsley G, Priston M, Williams-Lyn D, Sutherland J, Glaser T, Oliver E, Walter MA, Heathcote G, Levin A, Héon E: Phenotypic heterogeneity of CYP1B1:mutations in a patient with Peters' anomaly. J Med Genet 2001;38:324–326.
52 Mirzayans F, Pearce WG, MacDonald IM, Walter MA: Mutation of the PAX6 gene in patients with autosomal dominant keratitis. Am J Hum Genet 1995;57:539–548.
53 Bloch N: The different types of sclerocornea, their hereditary modes and concomitant congenital malformations (in French). J Genet Hum 1965;14:133–172.
54 Tahvanainen E, Villanueva AS, Forsius H, Salo P, de la Chapelle A: Dominantly and recessively inherited cornea plana congenita map to the same small region of chromosome 12. Genome Res 1996;6:249–254.
55 Pellegata NS, Dieguez-Lucena JL, Joensuu T, Lau S, Montgomery KT, Krahe R, Kivela T, Kucherlapati R, Forsius H, de la Chapelle A: Mutations in KERA, encoding keratocan, cause cornea plana. Nat Genet 2000;25:91–95.
56 Liu CY, Shiraishi A, Kao CW, Converse RL, Funderburgh JL, Corpuz LM, Conrad GW, Kao WW: The cloning of mouse keratocan cDNA and genomic DNA and the characterization of its expression during eye development. J Biol Chem 1998;273:22584–22588.
57 Edwards M, McGhee CN, Dean S: The genetics of keratoconus. Clin Exp Ophthalmol 2001;29:345–351.
58 Wang Y, Rabinowitz YS, Rotter JI, Yang H: Genetic epidemiological study of keratoconus: Evidence for major gene determination. Am J Med Genet 2000;93:403–409.
59 Bechara SJ, Waring GO, 3rd, Insler MS: Keratoconus in two pairs of identical twins. Cornea 1996;15:90–93.
60 Rabinowitz YS: Keratoconus. Surv Ophthalmol 1998;42:297–319.
61 Critchfield JW, Calandra AJ, Nesburn AB, Kenney MC: Keratoconus. I. Biochemical studies. Exp Eye Res 1988;46:953–963.
62 Kim WJ, Rabinowitz YS, Meisler DM, Wilson SE: Keratocyte apoptosis associated with keratoconus. Exp Eye Res 1999;69:475–481.
63 Collier SA, Madigan MC, Penfold PL: Expression of membrane-type 1 matrix metalloproteinase (MT1-MMP) and MMP-2 in normal and keratoconus corneas. Curr Eye Res 2000;21:662–668.
64 Cheng EL, Maruyama I, SundarRaj N, Sugar J, Feder RS, Yue BY: Expression of type XII collagen and hemidesmosome-associated proteins in keratoconus corneas. Curr Eye Res 2001; 22:333–340.
65 Ballabio A, Shapiro LJ: Steroid sulfatase deficiency and X-linked ichthyosis; in Scriver C, Beaudet A, Sly W, Valle D (eds): The Metabolic and Molecular Bases of Inherited Disease, ed 8. New York, McGraw-Hill, 2001, vol III, pp 4241–4262.
66 Scriver C, Beaudet A, Sly W, Valle D (eds): The Metabolic and Molecular Bases of Inherited Disease, ed 8. New York, McGraw-Hill, 2001, vol II, pp 2917–2918.

E. Héon, MD, FRCS(C)
Dept. of Ophthalmology and Vision Sciences
The Hospital for Sick Children
555 University Ave, Main Floor, Elm Wing, Rm. 165
Toronto, Ont M5G 1X8 (Canada)
Tel. 1 416 813 8606, Fax 1 416 813 8266, E-Mail ehcon@attglobal.net

Wissinger B, Kohl S, Langenbeck U (eds): Genetics in Ophthalmology.
Dev Ophthalmol. Basel, Karger, 2003, vol 37, pp 67–82

Molecular Genetics of Cataract

J. Fielding Hejtmancik, Nizar Smaoui

Ophthalmic Genetics and Visual Function Branch,
National Eye Institute, Bethesda, Md., USA

Abstract

Advances in genetic technology and analytical algorithms have greatly accelerated elucidation of the genetic contribution to cataractogenesis. Currently, 27 isolated or primary cataract loci have been identified by linkage analysis or mutational screening, and 20 are associated with specific genes. These are beginning to provide a framework for thinking of congenital cataracts. In addition to clues provided by the study of congenital and childhood cataracts, new experimental approaches to age-related cataracts are beginning to provide insights into its genetic origins.

Introduction

This chapter is intended to review briefly recent progress in delineating the molecular pathophysiology of inherited cataracts. Due to limitations of space, only a limited number of areas could be covered, and many worthwhile topics are discussed minimally, if at all. Among others, animal models of human cataractogenesis are only briefly touched on in a few examples. This topic has been reviewed recently [1]. In addition, while only minimal background information could be provided on lens biology, most of the basic science underlying identified mutations causing lens cataracts has been recently reviewed [2], and is accompanied by additional references for studies on inherited human cataracts as well. Finally, these same space limitations have required that all the work contributing to a topic cannot be referenced. In most cases, the most recent or primary reference is given, so that additional references can be obtained from the reference provided.

Cataract, defined here as a lens opacity, can have multiple causes and is generally associated with the breakdown of the lens microarchitecture. Vacuole formation will cause large fluctuations in density and hence abrupt changes in the index of refraction, resulting in light scattering. Light scattering and opacity also will occur if there are significant high-molecular-weight protein aggregates roughly 1,000 Å or more in size. The short-range ordered packing of the crystallins is important in this regard; to achieve and maintain lens transparency, crystallins must exist in a homogeneous phase. A variety of biochemical or physical insults can cause phase separation into protein-rich and protein-poor regions within the lens fibers, resulting in light scattering, and mutations in the crystallins can increase this susceptibility dramatically or themselves be sufficient to cause aggregation. The physical basis of lens transparency is beyond the scope of this chapter and is discussed and referenced elsewhere [2].

Cataracts have long been an interest of human geneticists. The CAE1 locus was initially linked with the Duffy blood group locus by Renwick and Lawler, and in 1968 the Duffy blood group locus was assigned to chromosome 1, making CAE1 the first human disease locus to be assigned to a human autosome. The likelihoods reported by Renwick were estimated using a computerized algorithm, the first to be analyzed in this fashion. The same program was used in the assignment of Duffy to chromosome 1. Since that time, with improvements in genotyping technologies and computational algorithms, there has been a dramatic acceleration in the pace of discovery of new Mendelian loci for both congenital and later onset cataracts, and attention is turning to age-related cataracts, which show a more complex multifactorial inheritance pattern. Finally, new techniques in molecular biology and molecular genetics are providing insights into metabolic and developmental pathways important in cataractogenesis and resistance of the lens to it.

Congenital Cataracts

Congenital cataracts are a significant cause of vision loss world wide, causing approximately one third of blindness in infants. Cataracts occur in approximately 0.01–0.06% of infants, and roughly half of congenital cataracts are hereditary. Cataracts can lead to permanent blindness by interfering with the sharp focus of light on the retina and resulting in failure to establish appropriate visual cortical synaptic connections with the retina. Prompt diagnosis and treatment can prevent this. Understanding the biology of the lens and the pathophysiology of selected types of cataract can yield insight into the process of cataractogenesis in general and provide a framework for the clinical approach to diagnosis and therapy.

Cataracts are known to occur in association with a large number of metabolic diseases and genetic syndromes [2]. Isolated congenital cataracts tend to be inherited in a Mendelian fashion with high penetrance, with autosomal dominant being more common than autosomal recessive. Because of this, and perhaps also because they are not lethal, congenital cataracts tend to be amenable to genetic mapping using linkage analysis. Currently, as listed in table 1, there are about 27 genetic loci to which isolated or primary cataracts have been mapped, although the number is constantly increasing. Of these, 8 are associated with additional abnormalities, mostly as part of developmental syndromes. These tend to result from mutations in genes encoding transcriptional activators, and most of these have been identified by sequencing candidate genes in patients with developmental anomalies. Two notable exceptions are the αB-crystallin gene, mutations in which can cause either isolated cataracts or cataracts associated with myopathy, and the ferritin gene, which causes the hyperferritinemia-cataract syndrome. While in some cases, e.g. some α- and β-crystallin mutations, inherited congenital cataracts are associated with microcornea and even microphakia, there are currently no identified developmental lesions causing isolated cataract.

Of the mapped loci for isolated congenital or infantile cataracts, 13 have been associated with mutations in specific genes. Of those families for whom the mutant gene is known, about half have mutations in crystallins, about a quarter have mutations in connexins, and the remainder are evenly split between aquaporin 0 (MIP), and the gene for the beaded filament protein BFSP2 (CP49 or phakenin). Inheritance of the same mutation in different families or even the same mutation within the same family can result in radically different cataract morphologies with widely varying implications for interference with vision, suggesting the importance of additional genes modifying the expression of the primary mutation associated with cataracts. Conversely, cataracts with similar or identical clinical presentations can result from mutations in quite different genes.

Mutations in the αA-crystallin gene have implicated both in autosomal recessive cataracts, which are associated with a chain termination mutation near the beginning of the protein [3], and in autosomal dominant cataracts, which are associated with nonconservative missense mutations [4]. The chain termination mutation would be expected to cause loss of function of the mutant protein, suggesting that half the normal levels of α-crystallin can provide sufficient chaperone-like activity and structural crystallin packing to establish and maintain lens transparency. These findings are consistent with data from knockout mice in which the αA-crystallin gene is disrupted. In mice lacking αA-crystallin the lenses are somewhat smaller in size and develop cataracts associated with the presence of inclusion bodies containing αB-crystallin, as well as increased amounts of αB-crystallin in the insoluble protein fraction of lens homogenates [5]. The occurrence of dominant cataracts

Table 1. Mapped human Mendelian cataract loci (in chromosomal order)

	Locus/gene[a]	Chromosome	Inh[b]	Morphology	Mutation	MIM	Ref.
1	**FOXE3**	1p32	AD	ASMD and cataracts	c.943–944insG (frameshift)	601094	19
2	**CCV** (Volkmann)	1p36	AD	Variable (progressive central and zonular nuclear cataract with sutural component)		115665	20
3	**CTPP** (posterior polar)	1p34–p36	AD	Posterior polar		116600	21
4	**GJA8** – connexin 50 (**CAE1, CZP1,** Duffy-linked)	1q21–q25	AD	Zonular pulverulent	P88S, E48K	116200	9
5	**CRYGC** – γC-crystallin (includes Coppock-like and variable nuclear)	2q33–q35	AD	Nuclear lamellar (Coppock-like), aculeiform, variable nuclear	T5P, c.117–118ins5bp (frameshift in first Greek key motif)	601286	2
6	**CRYGD** – γD-crystallin (includes **CACA**)	2q33–35	AD	Aculeiform, crystalline cataract	R14C, R37S, R58H	123690	22, 23
7		3p22–24.2	AR				24
8	**BFSP2 – CP49,** phakinin	3q21–q22	AD	Congenital nuclear and sutural cataracts in dln, juvenile lamellar cataracts in missense	delE233, R287W	603212	12
9	Ii blood group	6p23–p24	AR			110800	25
10	**EYA1 (ASD)**	8q13.3	AD	Congenital cataracts and anterior segment anomalies, 1 with BOR	R514G, E330K, G393S	601653	26
11	**CAAR**	9q13–q22	AR	Adult-onset pulverulent		212500	27
12	**SPG9** (spastic paraplegia with cataracts)	10q23.3–q24.2		Bilateral zonular cataracts		601162	28
13	***PITX3***	10q25		ASMD and cataracts	S13N, c.656–657ins17bp	602669	29

Table 1 (continued)

	Locus/gene[a]	Chromosome	Inh[b]	Morphology	Mutation	MIM	Ref.
14	**CRYAB** – αB-crystallin	11q22.3–33.1	AD	Posterior polar cataracts with del, or myopathy and cataracts	R120G, c.448delA (frameshift)	123590	30
15	**AQP0 (MIP, ADC)**	12q12–14.1	AD	Variable embryonal nuclear, progressive bilateral punctuate, with asymmetric polar opacification	E134G, T138R, delG213	601286	11
16	**GJA3** – connexin 46 **(CZP3, CAE3)**	13q11–q13	AD	Zonular pulverulent	N63S, P187L, c.1136–1137insC (frameshift)	601885	10
17	**CCPSO**	15q21–q22	AD	Central pouch-like with sutural opacities		605728	31
18	**CAM** (Marner)	16q22	AD	Variable (progressive central and zonular nuclear, anterior polar or stellate)		116800	32
19	**MAF**	16q22	AD	Cataract, iris coloboma, microcornea	R288P	177074	33
20	**CTAA2** (anterior polar)	17p13	AD	Anterior polar		601202	34
21	**CRYBA1** – βA3-crystallin **(CCZS)**	17q11–q12	AD	Nuclear lamellar with sutural component	IVS3+1G>A, IVS3+1G>C	600881	35
22	**CCA1** (cerulean – blue dot)	17q24	AD	Cerulean (nuclear and cortical)		115660	36
23	**FTL** – ferritin (hyperferritinemia and cataracts)	19q13.4	AD			154045	37
24	**CPP3**	20p12–q12	AD	Progressive, disc-shaped posterior subcapsular opacity		605387	38

Table 1 (continued)

	Locus/gene[a]	Chromosome	Inh[b]	Morphology	Mutation	MIM	Ref.
25	**CRYAA** – αA-crystallin	21q22.3	AD, AR	Congenital zonular nuclear with cortical and posterior subcapsular as adults	R116C, W9X (AR), R49C	123580	3, 4
26	**CRYBB2** – βB2-crystallin (**CCA2**, cerulean – blue dot)	22q11.2	AD	Cerulean, Coppock-like (CCL)	Q155X (both cerulean and CCL), gene conversion	601547	39
27	X-linked cataracts	Xp	XL	Possibly allelic with Nance-Horan syndrome		302350	40

[a]Gene symbols shown in bold and disease loci in italics. Gene/loci synonyms and type of cataract in parentheses.
[b]Inh = Inheritance.

with the missense mutations suggests that the mutant αA-crystallin protein exerts a deleterious effect that actively damages the lens cell or its constituent proteins, or inhibits the function of the remaining normal α-crystallin, rather than acting through loss of chaperone function as the recessive cataract appears to do.

Because αA- and αB-crystallin are found in the lens associated into large multimeric complexes and function similarly in vitro, one might expect that mutations in αB-crystallin would have a similar effect to those in αA-crystallin, at least in the lens. However, the first human mutation reported in αB-crystallin was associated with desmin-related myopathy and only discrete cataracts. This was a missense mutation that reduced αB-crystallin chaperone activity dramatically, causing aggregation and precipitation of the protein under stress. The myopathy associated with this mutation is probably related to the expression of αB-crystallin, but not αA-crystallin, in muscle cells, where it binds and presumably stabilizes desmin. Similarly, an αB-crystallin knockout mouse exhibits myopathy without cataracts [6]. In contrast, a deletion in the αB-crystallin gene causing a frameshift and expression of 184 aberrant amino acids causes autosomal dominant cataracts not associated with myopathy. This seems more similar to the dominant αA-crystallin-associated cataract, with the aberrant protein likely to have a toxic effect on the lens cells.

Most mutations described in the βγ-crystallins would be expected to cause gross abnormalities in the protein structure, presumably resulting in an unstable protein that precipitates from solution and serves as a nidus for additional protein denaturation and precipitation, eventually resulting in cataract formation. These include missense mutations, insertions changing the reading frame and causing expression of aberrant peptides with premature termination, and splice mutations as shown in table 1. Mutations in the γ-crystallins tend to produce nuclear or zonular cataracts, consistent with their high level of expression in the lens nucleus, although the phenotypes can vary significantly. The cataract phenotypes reported with mutations in the β-crystallins is somewhat more varied, ranging in different families from zonular pulverulent with or without involvement of the sutures to cerulean cataracts. The association of identical mutations in βB2-crystallin in different families with nuclear lamellar Coppock-like and cerulean cataracts emphasizes the importance of modifying genes in the phenotypic expression of these mutations.

Recently, two mutations in γD-crystallin, R36S and R58H, have been shown not to alter the protein fold, but rather to alter the surface characteristics of the protein [7]. This, in turn, lowers the solubility and enhances the crystal nucleation rate of these mutants so that they precipitate out of solution, in at least one case actually forming crystals in the lens. In a third mutation in γD-crystallin, R14C, the protein also maintains a normal protein fold, but is susceptible to thiol-mediated aggregation [8]. These results emphasize that crystallins need not undergo denaturation or other major changes in their protein folds to cause cataracts.

The hyperferritinemia-cataract syndrome is a recently described disorder in which cataracts are associated with hyperferritinemia without iron overload. Ferritin L levels in the lens can increase dramatically. The molecular pathology lies in the Ferritin L iron-responsive element, a stem loop structure in the 5′-untranslated region of the ferritin mRNA. Normally, this structure binds a cytoplasmic protein, the iron regulatory protein, which then inhibits translation of ferritin mRNA, which may exist in the lens at levels approaching that of a lens crystallin. Mutation of this structure and overexpression of ferritin by loss of translational control in the hyperferritinemia-cataract syndrome results in crystallization of ferritin in the lens, and other tissues as well, with cataracts resulting in a fashion similar to that of the γ-crystallin mutations described above and appearing as breadcrumb-like opacities in the cortex and nucleus. Presence at such high levels in the protein-rich lens cytoplasm requires that crystallins or other proteins must be exceptionally soluble.

Connexins 46 and 50 are constituents of gap junctions, on which the avascular lens depends for nutrition and intercellular communication. At least one cataract-associated mutation in the connexin 50 gene, the P88S missense mutation in the second transmembrane domain, has been shown to result in a connexin that fails to form functional gap junctional channels, and of which incorporation of even a single protein into a gap junction inhibits channel function in *Xenopus* oocytes [9]. Mutant connexin 46 proteins are also associated with cataracts. The N63S missense mutation in the first extracellular domain, and a frameshift mutation at residue 380 causing read-through into the 3′-untranslated region until an in-frame stop codon 90 nucleotides downstream from the wild-type stop codon, also fail to form intercellular channels in paired *Xenopus* oocytes [10]. However, they were unable to participate in gap junction formation at all, and thus did not inhibit channel function by products of the normal gene. Mutations in both connexin 46 and connexin 50 produce phenotypically similar autosomal dominant zonular pulverulent cataracts.

Lamellar and polymorphic cataracts have been associated with missense mutations in the MIP gene. One mutation, E134G, is associated with a nonprogressive congenital lamellar cataract, and the second T138R is associated with multifocal opacities that increase in severity throughout life. When expressed in *Xenopus laevis* oocytes, both of these mutations appear to act by interfering with normal trafficking of MIP to the plasma membrane and thus with water channel activity [11]. In addition, both mutant proteins appear to interfere with water channel activity by normal MIP, consistent with the autosomal dominant inheritance of the cataracts.

Beaded filaments are a type of intermediate filament unique to the lens fiber cells. They are made up of bfsp1 (also called CP115 or filensin) and bfsp2 (also called CP49 or phakinin), highly divergent intermediate filament proteins that combine in the presence of α-crystallin to form the appropriate beaded structure. Cataracts in three families have been mapped to regions including the beaded filament structural protein bfsp2, or cp49. In one family the cataracts are associated with a nonconservative missense mutation in exon 4 substituting a tryptophan for an evolutionarily conserved arginine in the central rod domain of the protein [12]. A deletion resulting in loss of glu233 in this protein has also been associated with cataracts [13]. These cataracts are nuclear or nuclear lamellar, with some involvement of the sutures, consistent with fiber cell-specific expression of the beaded filament proteins.

Congenital cataracts interfere with vision at a time when neuronal connections are still being formed in the visual cortex and optic tracts. Because of this, they can interfere with the formation of neuronal connections necessary for visual processing, resulting in amblyopia. This can be prevented or at least ameliorated by a combination of early surgery in the first eye rapidly followed by surgery on the second eye. In addition, bilateral occlusion between the two operations, careful postoperative monitoring, and early correction of aphakia also seem to provide a better long-term outcome. While some differences in approach remain, there is a general consensus that aggressive therapy is indicated, especially in otherwise uncomplicated unilateral cataracts with parents highly motivated to follow rigorous occlusion protocols.

Age-Related Cataracts

While congenital cataracts can be particularly threatening to vision and up to one half of all congenital cataracts are inherited, they affect relatively few individuals in comparison to age-related cataracts, which are responsible for just under half of all blindness worldwide. Cataract surgery is the most frequently performed surgical procedure in the USA, and because of its demographics, it has been estimated that delaying the development of cataract by 10 years would decrease the need for cataract surgery by about 45%. Age-related cataract is associated with a number of environmental risk factors, including cigarette smoking or chronic exposure to wood smoke, obesity or elevated blood glucose levels, poor infantile growth, exposure to ultraviolet light, and alcohol consumption. Conversely, antioxidant vitamins seem to have a protective effect.

Obviously, in age-related cataracts, the lens develops at least reasonably normally during infancy and remains clear in childhood. Then, by somewhat arbitrary definition, at some time after 40 years of age, progressive opacities begin to form in the lens. As mentioned above, these opacities almost certainly result at least in part from the cumulative damage of environmental insults on lens proteins and cells. Lens proteins are known to undergo a wide variety of alterations with age, and many of these are accelerated in the presence of oxidative, osmotic, or other stresses, which are also known to be associated with cataracts. In the case of lens crystallins, these include proteolysis, an increase in disulfide bridges, deamidation of asparagine and glutamine residues, racemization of aspartic acid residues, phosphorylation, nonenzymatic glycosylation and carbamylation. Many of these changes have been found to be increased in cataractous lenses and to be induced in vitro or in model systems by the same stresses epidemiologically associated with cataracts.

The lens crystallins form one obvious target for this accumulated damage, although they are certainly not the only one. Thus, as the β- and γ-crystallins slowly accumulate damage over the lifetime of an individual, they would loose the ability to participate in appropriate intermolecular interactions, and even to remain in solution. As these crystallins begin to denature and precipitate, they are bound by the α-crystallins, which have a chaperone-like activity. Binding by α-crystallins maintains solubility of βγ-crystallins and reduces light scattering, but in general, the α-crystallins appear not to renature their target proteins and release them into the cytoplasm, as do true chaperones. Rather, they hold them in complexes that, while soluble, increase in size as additional damaged protein is bound over time until they themselves begin to approach sizes sufficient to scatter light. Eventually, it seems likely that the available α-crystallin is overwhelmed by increasing amounts of modified βγ-crystallin and the complexes precipitate within the lens cell, forming the insoluble protein fraction that is known to increase with age and in cataractous lenses.

Whether proteins in the insoluble fraction become insoluble upon complete or partial denaturation, as would be implied by the schema above, or whether they simply become less soluble due to modifications that leave their protein folds largely intact, is not currently known. Classically, it was believed that insoluble proteins in the lens became insoluble because they were denatured. There is a large body of data showing that insoluble protein in the aged cataractous lens not only is denatured and cross-linked, but that a fraction exists as relatively short peptides cleaved from larger proteins. There are even suggestions that this denatured protein exists as amyloid, although it would, at least initially, be intracellular, and there is little evidence that it causes precipitation of normal protein from the lens fiber cell cytoplasm. However, it seems clear that the presence of large amounts of unstable or precipitated crystallin, or other protein, does damage to the lens cell and eventually contributes to cataracts not only directly through light scattering by protein aggregates but eventually also through disruption of cellular architecture. This is clear from numerous mouse models of cataracts resulting from crystallin mutations.

Genetic Epidemiology of Age-Related Cataracts

There is increasing evidence that genetic factors are important in the pathogenesis of age-related cataract [14]. In 1991, the Lens Opacity Case Control Study indicated that a positive family history was a risk factor for mixed nuclear and cortical cataracts, and the Italian-American Eye Study supported a similar role for family history as a risk factor in cortical, mixed nuclear and cortical, and posterior subcapsular cataracts. In 1994, the Framingham Offspring Eye Study showed that individuals with an affected sibling had three times the likelihood of also having a cataract. In 1993 and 1995, the Beaver Dam Eye Study examined nuclear sclerotic cataracts using sibling correlations and segregation analysis. While a random environmental major effect was rejected by this study, Mendelian transmission was not rejected, and the results suggested that a single major gene could account for as much as 35% of nuclear and up to 75% of cortical cataract variability. Most recently, in 2001 the twin eye study demonstrated significant genetic influence of age-related cortical cataract, with heritability accounting for 53–58% of the liability for age-related cortical cataract. This hereditary tendency was consistent with a combination of additive and dominant genes, with dominant genes accounting for 38–53% of the genetic effect, depending on whether cataracts were scored using the Oxford or Wilmer grading systems. Similarly, genetic factors were found to account for approximately 48% of the risk for nuclear cataract.

In addition to epidemiological evidence implicating genetic factors in age-related cataract, a number of inherited cataracts with post-infantile age of onset or progression of the opacity throughout life have been described. Mutations in BFSP2 can cause juvenile cataracts, the Marner and Volkmann cataracts can be progressive, mutations in aquaporin 0 (MIP) and γC-crystallin can cause progressive cataracts, and the CAAR locus is linked to familial adult-onset pulverulent cataracts. These all suggest that for at least some genes, a mutation that severely disrupts the protein or inhibits its function might result in congenital cataracts inherited in a highly penetrant Mendelian fashion, while a mutation that causes less severe damage to the same protein or impairs its function only mildly might contribute to age-related cataracts in a more complex multifactorial fashion. Similarly, mutations that severely disrupt the lens cell architecture or environment might produce congenital cataracts, while others that cause relatively mild disruption of lens cell homeostasis might contribute to age-related cataract.

Galactosemic cataracts provide an interesting example of this principle. Deficiencies of galactokinase, galactose-1-phosphate uridyl transferase, and severe deficiencies of uridine diphosphate 1–4 epimerase cause cataracts as a result of galactitol accumulation and subsequent osmotic swelling. The latter two are also associated with vomiting, failure to thrive, liver disease, and mental retardation if untreated, while the cataracts in galactokinase deficiency are isolated. Interestingly, galactosemic cataracts initially are reversible both in human patients and in animal models. In 2001, a novel variant of galactokinase, the Osaka variant with an A198V substitution, was shown to be associated with a significant increase in bilateral cataracts in adults [15]. It results in instability of the mutant protein and is responsible for mild galactokinase deficiency leaving about 20% of normal levels. This variant allele frequency occurs in 4.1% in Japanese overall and 7.1% of Japanese with cataracts. The allele was also present in 2.8% of Koreans but had a lower incidence in Chinese and was not seen in blacks or whites from the USA.

This result fits in well with the known influence of hyperglycemia on age-related cataract. That these cataracts result from polyol accumulation is suggested by work in galactosemic dogs and transgenic and knockout mice. Dogs have aldose reductase levels similar to those in humans and when stressed readily develop sugar cataracts that are prevented by aldose reductase inhibitors. Mice, which have very low aldose reductase activity in the lens, are naturally resistant to sugar cataracts, either galactosemic or hyperglycemic. However, upon transgenic expression of aldose reductase, mice readily develop cataracts, especially when galactokinase or sorbitol dehydrogenase is deleted. Consistent with these animal data are the recent findings that susceptibility to both cataracts and retinopathy as a diabetic complications in humans is associated with specific allele Z of the microsatellite polymorphism at 5′ of the aldose reductase gene.

As mentioned above, the epidemiology of cataracts strongly implicates oxidative stress, and especially photo-oxidation, as risk factors for age-related cataracts. This suggests that key enzymes of metabolic pathways maintaining the reducing environment of the lens might be candidates for involvement in age-related cataracts. One such candidate that has shown inconclusive results is glutathione S-transferase, with studies showing increased, unchanged, and decreased risk for age-related cataracts with the null allele of glutathione S-transferase M in different populations. It is possible that glutathione S-transferase P might show a stronger effect on risk of age-related cataracts, since it is the most prevalent glutathione S-transferase in the lens. Thiol transferase (glutaredoxin) has been shown to increase in response to oxidative stress in immortalized human lens epithelial cells [16], and would also represent a reasonable gene for consideration. As with glutathione S-transferase, however, this might be complicated by the occurrence of more than one form in the lens.

Experimental Approaches to Age-Related Cataracts

In addition to the genetic epidemiological studies of age-related cataracts, a number of experimental approaches have provided insight into the genetics of age-related cataract. One approach has been to identify mRNAs that show a substantial increase or decrease in cataractous lenses [17]. Thus, this approach does not directly identify genes that, when mutated, cause or contribute to cataract, as do the more direct genetic studies described above. Indeed, it will not identify genes with missense mutations at all, unless the mutation actually inhibits translation and thereby destabilizes the mRNA. In addition, mRNA levels tend to be measured in lens epithelia cells, while in age-related cataracts the opacities tend to occur in the nuclear or cortical fiber cells. Rather, it identifies genes that belong to metabolic or regulatory networks that are activated or inhibited as a result of cataract formation. These genes may or may not contribute to cataractogenesis in the lenses in which they were identified. However, it is reasonable to hypothesize that if cataracts or the stresses that cause them induce expression of these genes, mutations in these genes might then contribute to cataract formation in a lens subjected to those same types of stress.

Thus, genes identified in this fashion are certainly candidates for the more direct genetic analysis described above.

The genes identified in this fashion form an interesting and rather surprising group. Since most of these genes have been identified by differential display RT-PCR, they do not represent an exhaustive catalogue of transcriptional changes in cataractous lenses. The mRNAs encoding metallothionine IIa and osteonectin (also known as SPARC, secreted acidic protein rich in cysteines) are increased, while those for protein phosphatase 2A regulatory subunit and some ribosomal proteins including L21, L15, L13a and L7a are decreased. These findings are consistent with induction of metallothionine by lens epithelia in the face of oxidative and perhaps toxic stress in the presence of divalent cations, and also with a protective role for SPARC. Although the function of SPARC, which also binds calcium, is less well understood, it is known to increase in the face of cellular injury and to be involved in cellular growth and growth factor control. Similarly, the decrease in protein phosphatase 2A regulatory subunit is consistent with decreasing cell division in the lens epithelia, while decreasing ribosomal proteins would be expected in the face of the corresponding decreases in protein synthesis. That these reactive proteins might cause cataracts if aberrantly expressed is supported by the occurrence of cataracts in mice lacking SPARC.

Another approach to identifying genes belonging to regulatory or metabolic pathways that might be important in age-related cataracts has been to examine mRNAs whose expression is modified in lens cells subjected to oxidative stress [18]. As described above, a large body of experimental and epidemiological evidence implicates oxidative, and especially photo-oxidative, stress in age-related cataract. When αTN4 lens cells transformed with SV40 t antigen are exposed to increasing levels of hydrogen peroxide, they adapt by increasing expression of a limited number of genes. Among the genes identified by differential display, RT-PCR are predominantly antioxidant and cellular defense enzymes including catalase, which increased 14-fold. Reticulocalbin increased 6-fold, while glutathione peroxidase, ferritin and αB-crystallin each increased 2-fold. αA-crystallin mRNA levels decreased to one fifth of baseline, while mRNAs for aldose reductase and mitochondrial enzymes showed no change. While the relationship of these genes to age-related cataract is somewhat less certain, they are logical candidates based on their known roles in oxidative stress.

Conclusions and Future Prospects

In summary, significant inroads are being made into understanding the genetics of human congenital cataracts, and the first initial insights are opening up for age-related cataracts. It has been estimated that there might be as many as 40 genes contributing to congenital cataracts in the mouse, and it would be reasonable to assume a similar number in humans. As our understanding of congenital and age-related cataracts increases, the relationship between their genetic causes becomes correspondingly more approachable. This is important for the study of age-related cataracts, because delineation of their genetics is much more difficult, due to their complex inheritance and late onset. Understanding the genetics of these cataracts is of paramount importance in order to guide development of a medical therapy that will prevent or delay their onset, lessening their burden on the aging population and the requirement for large numbers of surgical procedures.

References

1 Graw J: Mouse models of congenital cataract. Eye 1999;13:438–444.

2 Hejtmancik JF, Kaiser-Kupfer MI, Piatigorsky J: Molecular biology and inherited disorders of the eye lens; in Scriver CR, Beaudet AL, Valle D, Sly WS, Childs B, Kinzler KW, Vogelstein B (eds): The Metabolic and Molecular Basis of Inherited Disease, 8 ed. New York, McGraw-Hill, 2001, vol 241, pp 6033–6062.

3 Pras E, Frydman M, Levy-Nissenbaum E, Bakhan T, Raz J, Assia EI, Goldman B, Pras E: A nonsense mutation (W9X) in CRYAA causes autosomal recessive cataract in an inbred Jewish Persian family. Invest Ophthalmol Vis Sci 2000;41:3511–3515.

4 Litt M, Kramer P, LaMorticella DM, Murphey W, Lovrien EW, Weleber RG: Autosomal dominant congenital cataract associated with a missense mutation in the human α-crystallin gene CRYAA. Hum Mol Genet 1998;7:471–474.

5 Brady JP, Garland D, Duglas-Tabor Y, Robison WG Jr, Groome A, Wawrousek EF: Targeted disruption of the mouse αA-crystallin gene induces cataract and cytoplasmic inclusion bodies containing the small heat-shock protein αB-crystallin. Proc Natl Acad Sci USA 1997;94:884–889.

6 Brady JP, Garland DL, Green DE, Tamm ER, Giblin FJ, Wawrousek EF: αB-crystallin in lens development and muscle integrity: A gene knockout approach. Invest Ophthalmol Vis Sci 2001;42:2924–2934.

7 Pande A, Pande J, Asherie N, Lomakin A, Ogun O, King J, Benedek GB: Crystal cataracts: Human genetic cataract caused by protein crystallization. Proc Natl Acad Sci USA 2001;98:6116–6120.

8 Pande A, Pande J, Asherie N, Lomakin A, Ogun O, King JA, Lubsen NH, Walton D, Benedek GB: Molecular basis of a progressive juvenile-onset hereditary cataract. Proc Natl Acad Sci USA 2000;97:1993–1998.

9 Pal JD, Berthoud VM, Beyer EC, Mackay D, Shiels A, Ebihara L: Molecular mechanism underlying a Cx50-linked congenital cataract. Am J Physiol 1999;276:C1443–C1446.

10 Pal JD, Liu X, Mackay D, Shiels A, Berthoud VM, Beyer EC, Ebihara L: Connexin46 mutations linked to congenital cataract show loss of gap junction channel function. Am J Physiol 2000;279:C596–C602.

11 Francis P, Chung JJ, Yasui M, Berry V, Moore A, Wyatt MK, Wistow G, Bhattacharya SS, Agre P: Functional impairment of lens aquaporin in two families with dominantly inherited cataracts. Hum Mol Genet 2000;9:2329–2334.

12 Conley YP, Erturk D, Keverline A, Mah TS, Keravala A, Barnes LR, Bruchis A, Hess JF, FitzGerald PG, Weeks DE, Ferrell RE, Gorin MB: A juvenile-onset, progressive cataract locus on chromosome 3q21–q22 is associated with a missense mutation in the beaded filament structural protein-2. Am J Hum Genet 2000;66:1426–1431.

13 Jakobs PM, Hess JF, FitzGerald PG, Kramer P, Weleber RG, Litt M: Autosomal-dominant congenital cataract associated with a deletion mutation in the human beaded filament protein gene BFSP2. Am J Hum Genet 2000;66:1432–1436.
14 McCarty CA, Taylor HR: The genetics of cataract. Invest Ophthalmol Vis Sci 2001;42:1677–1678.
15 Okano Y, Asada M, Fujimoto A, Ohtake A, Murayama K, Hsiao KJ, Choeh K, Yang Y, Cao Q, Reichardt JK, Niihira S, Imamura T, Yamano T: A genetic factor for age-related cataract: Identification and characterization of a novel galactokinase variant, 'Osaka,' in Asians. Am J Hum Genet 2001;68:1036–1042.
16 Raghavachari N, Krysan K, Xing K, Lou MF: Regulation of thiol transferase expression in human lens epithelial cells. Invest Ophthalmol Vis Sci 2001;42:1002–1008.
17 Zhang W, Hawse J, Huang Q, Sheets N, Miller KM, Horwitz J, Kantorow M: Decreased expression of ribosomal proteins in human age-related cataract. Invest Ophthalmol Vis Sci 2002;43: 198–204.
18 Carper D, John M, Chen Z, Subramanian S, Wang R, Ma W, Spector A: Gene expression analysis of an H(2)O(2)-resistant lens epithelial cell line. Free Radic Biol Med 2001;31:90–97.
19 Semina EV, Brownell I, Mintz-Hittner HA, Murray JC, Jamrich M: Mutations in the human forkhead transcription factor FOXE3 associated with anterior segment ocular dysgenesis and cataracts. Hum Mol Genet 2001;10:231–236.
20 Eiberg H, Lund AM, Warburg M, Rosenberg T: Assignment of congenital cataract Volkmann type (CCV) to chromosome 1p36. Hum Genet 1995;96:33–38.
21 Ionides AC, Berry V, Mackay DS, Moore AT, Bhattacharya SS, Shiels A: A locus for autosomal dominant posterior polar cataract on chromosome 1p. Hum Mol Genet 1997;6:47–51.
22 Héon E, Priston M, Schorderet DF, Billingsley GD, Girard PO, Lubsen N, Munier FL: The γ-crystallins and human cataracts: A puzzle made clearer. Am J Hum Genet 1999;65:1261–1267.
23 Kmoch S, Brynda J, Asfaw B, Bezouska K, Novak P, Rezacova P, Ondrova L, Filipec M, Sedlacek J, Elleder M: Link between a novel human γD-crystallin allele and a unique cataract phenotype explained by protein crystallography. Hum Mol Genet 2000;9:1779–1786.
24 Pras E, Pras E, Bakhan T, Levy-Nissenbaum E, Lahat H, Assia EI, Garzozi HJ, Kastner DL, Goldman B, Frydman M: A gene causing autosomal recessive cataract maps to the short arm of chromosome 3. Isr Med Assoc J 2001;3:559–562.
25 Yamaguchi H, Okubo Y, Tanaka M: A note on possible close linkage between the Ii blood locus and a congenital cataract locus. Proc Jpn Acad 1972;48:625–628.
26 Azuma N, Hirakiyama A, Inoue T, Asaka A, Yamada M: Mutations of a human homologue of the *Drosophila* eyes absent gene (EYA1) detected in patients with congenital cataracts and ocular anterior segment anomalies. Hum Mol Genet 2000;9:363–366.
27 Héon E, Paterson AD, Fraser M, Billingsley G, Priston M, Balmer A, Schorderet DF, Verner A, Hudson TJ, Munier FL: A progressive autosomal recessive cataract locus maps to chromosome 9q13–q22. Am J Hum Genet 2001;68:772–777.
28 Seri M, Cusano R, Forabosco P, Cinti R, Caroli F, Picco P, Bini R, Morra VB, DeMichele G, Lerone M, Silengo M, Pela I, Borrone C, Romeo G, Devoto M: Genetic mapping to 10q23.3–q24.2, in a large Italian pedigree, of a new syndrome showing bilateral cataracts, gastroesophageal reflux, and spastic paraparesis with amyotrophy. Am J Hum Genet 1999; 64:586–593.
29 Semina EV, Ferrell RE, Mintz-Hittner HA, Bitoun P, Alward WL, Reiter RS, Funkhauser C, Daack-Hirsch S, Murray JC: A novel homeobox gene PITX3 is mutated in families with autosomal-dominant cataracts and ASMD. Nat Genet 1998;19:167–170.
30 Berry V, Francis P, Reddy MA, Collyer D, Vithana E, MacKay I, Dawson G, Carey AH, Moore A, Bhattacharya SS, Quinlan RA: αB-crystallin gene (CRYAB) mutation causes dominant congenital posterior polar cataract in humans. Am J Hum Genet 2001;69:1141–1145.
31 Vanita, Singh JR, Sarhadi VK, Singh D, Reis A, Rueschendorf F, Becker-Follmann J, Jung M, Sperling K: A Novel Form of 'Central Pouchlike' Cataract, with Sutural Opacities, Maps to Chromosome 15q21–22. Am J Hum Genet 2000;68:509–514.
32 Eiberg H, Marner E, Rosenberg T, Mohr J: Marner's cataract (CAM) assigned to chromosome 16: Linkage to haptoglobin. Clin Genet 1988;34:272–275.

33 Jamieson RV, Perveen R, Kerr B, Carette M, Yardley J, Héon E, Wirth MG, van Heyningen V, Donnai D, Munier F, Black GC: Domain disruption and mutation of the bZIP transcription factor, MAF, associated with cataract, ocular anterior segment dysgenesis and coloboma. Hum Mol Genet 2002;11:33–42.
34 Berry V, Ionides AC, Moore AT, Plant C, Bhattacharya SS, Shiels A: A locus for autosomal dominant anterior polar cataract on chromosome 17p. Hum Mol Genet 1996;5:415–419.
35 Kannabiran C, Rogan PK, Olmos L, Basti S, Rao GN, Kaiser-Kupfer M, Hejtmancik JF: Autosomal dominant zonular cataract with sutural opacities is associated with a splice site mutation in the βA3/A1-crystallin gene. Mol Vis 1998;4:21.
36 Armitage MM, Kivlin JD, Ferrell RE: A progressive early-onset cataract gene maps to human chromosome 17q24. Nat Genet 1995;9:37–40.
37 Beaumont C, Leneuve P, Devaux I, Scoazec JY, Berthier M, Loiseau MN, Grandchamp B, Bonneau D: Mutation in the iron responsive element of the L ferritin mRNA in a family with dominant hyperferritinaemia and cataract. Nat Genet 1995;11:444–446.
38 Yamada K, Tomita H, Yoshiura K, Kondo S, Wakui K, Fukushima Y, Ikegawa S, Nakamura Y, Amemiya T, Niikawa N: An autosomal dominant posterior polar cataract locus maps to human chromosome 20p12–q12. Eur J Hum Genet 2000;8:535–539.
39 Litt M, Carrero-Valenzuela R, LaMorticella DM, Schultz DW, Mitchell TN, Kramer P, Maumenee IH: Autosomal dominant cerulean cataract is associated with a chain termination mutation in the human β-crystallin gene CRYBB2. Hum Mol Genet 1997;6:665–668.
40 Francis PJ, Berry V, Hardcastle AJ, Maher ER, Moore AT, Bhattacharya SS: A locus for isolated cataract on human Xp. J Med Genet 2002;39:105–109.

James Fielding Hejtmancik, MD, PhD
OGVFB/NEI/NIH, Building 10, Room 10B10,
10 Center DR MSC 1860, Bethesda, MD 20892–1860 (USA)
Tel. +1 301 4968300, Fax +1 301 4351598, E-Mail f3h@helix.nih.gov

Wissinger B, Kohl S, Langenbeck U (eds): Genetics in Ophthalmology.
Dev Ophthalmol. Basel, Karger, 2003, vol 37, pp 83–93

Progress in the Genetics of Glaucoma

N. Weisschuh[a], *U. Schiefer*[b]

[a] Molekulargenetisches Labor Universitäts-Augenklinik Tübingen und
[b] Abteilung Pathophysiologie des Sehens und Neuro-Ophthalmologie, Universitäts-Augenklinik Tübingen, Deutschland

Abstract

The term glaucoma describes a heterogeneous group of optic neuropathies that lead to optic nerve atrophy and permanent loss of vision. It is the second most prevalent cause of bilateral blindness in the Western world and affects over 60 million people worldwide. The hereditary forms of glaucoma are genetically heterogeneous. Different forms of glaucoma can be distinguished: the primary open-angle glaucoma of adult onset is the most common, representing approximately half of all cases. The juvenile-onset open-angle glaucoma is an uncommon autosomal dominant form of glaucoma with manifestation predominantly before the fourth decade of life. The primary congenital glaucoma is a clinical and genetic entity clearly distinct from the juvenile form, following an autosomal recessive mode of inheritance. At least eight loci have been linked to glaucoma (*GLC1A-F, GLC3A/B*) and three genes have been identified to date: *MYOC*, *CYP1B1* and *OPTN*. In the last decade, there has been much progress in finding new genes, detecting disease-related mutations and determining allele frequencies within populations of different ethnical backgrounds, but little is known about the function of the mutated gene products and the underlying pathogenic mechanisms. This chapter attempts to summarize the current knowledge regarding glaucoma-associated genes.

Introduction

Glaucoma has a high socio-economic impact as it is the second most prevalent cause of bilateral blindness in the Western world, after cataracts. The prevalence of glaucoma worldwide was estimated to be about 67 million subjects in the year 2000 [1]. Since glaucoma rates rise exponentially with age, more glaucoma cases are expected in the future as an effect of the ageing Western populations. The disease is insidious and affected patients frequently have no symptoms, especially in the early stages. When detected early, most cases can be successfully treated with medications, laser treatment or surgery.

Clinical Criteria for Glaucoma

The term glaucoma defines a heterogeneous group of eye disorders characterized by progressive excavation of the optic nerve head with associated visual field losses. Finding an appropriate definition based on fixed values concerning the structural and functional damages that occur in glaucoma is rather difficult. Foster et al. [2] describe a scheme for diagnosis of glaucoma which is based on broadly accepted principles. The diagnosis is made according to three levels of evidence that refer to the vertical cup:disc ratio and characteristics of glaucomatous field defects.

Although the level of intraocular pressure (IOP) is one of the major risk factors for developing glaucoma, it is not used a prime criteria as not all cases of glaucoma involve abnormally high levels of IOP and, conversely, people with statistically elevated IOP may show no evidence of optic neuropathy.

The underlying pathogenic mechanisms of glaucoma are not quite understood. Yet the loss of vision in all subtypes is due to the gradual death of retinal ganglion cells, leading to the atrophy of the optic nerve. This may be due to an increased resistance to the outflow of aqueous humour which fills the anterior chamber of the eye. This fluid is continuously produced by the ciliary body and exits the eye through the trabecular meshwork.

Phenotypes

The most common form of glaucoma is the open-angle glaucoma (OAG) representing approximately half of all cases [3]. Affected patients show optic nerve damage but no evidence of angle closure on gonioscopy. Patients with angle-closure glaucoma (ACG) show optic nerve damage, occludable drainage angles and obstruction of the trabecular meshwork [2]. ACG compared to OAG is reported to be less common among Europeans [3] in contrast to some Asian populations where it is more prevalent [4].

Both OAG and ACG can be primary or secondary. A primary condition is one that cannot be attributed to any known cause. A secondary condition can be traced to another cause, such as previous injury or illness. Primary forms of glaucoma are typically bilateral and influenced by genetic determinants.

Those subtypes of glaucoma that were shown to be associated with susceptibility loci are specified in the following.

Table 1. Glaucoma-associated genes

Gene	Mapped Loci	Protein	Phenotypes (and inheritance)
MYOC	GLC1A Chromosome 1q23–24	Myocilin	JOAG (autosomal dominant), POAG
CYP1B1	GLC3A Chromosome 2p21	Cytochrome P1B1	PCG (autosomal recessive)
OPTN	GLC1E Chromosome 10p14–p15	Optineurin	POAG
PITX2	RIEG1, IRID2 Chromosome 4q25–27	Homeobox transcription factor	Rieger syndrome, iris hypoplasia
FOXC1	Chromosome 6p25	Forkhead transcription factor	PCG, Rieger anomaly, Axenfeld anomaly, iris hypoplasia
	GLC1B on chromosome 2cen–q13		POAG
	GLC1C on chromosome 3q21–q24		POAG
	GLC1D on chromosome 8q23		POAG
	GLC1F on chromosome 7q35–q36		POAG
	GLC3B on chromosome 1p36		PCG
	RIEG2 on chromosome 13q14		Rieger syndrome

Primary OAG (POAG) has varying prevalence in different populations. With increasing age the frequency rises: including its preliminary stages, POAG affects about 1–2% of the general population over 40. Beyond the age of 75 the frequency rises to 7–8% [3]. The prevalence of POAG in first-degree relatives of affected patients has been documented to be 7–10 times higher than that of the general population [5, 6]. It has also been shown that POAG is more prevalent among black people [3]. Although the heredity of POAG is high, a simple mode of inheritance is not apparent. Instead it is assumed to be inherited as a complex trait without an obvious segregation pattern.

Normal tension glaucoma (NTG) is an important subtype of POAG, accounting for approximately 20–50% of all POAG cases [4, 7]. Patients with NTG show IOPs that are within the statistical normal range of the population (10–20 mm Hg).

A rare form of OAG is the POAG of juvenile-onset OAG (JOAG). Unlike the adult-onset disease, the juvenile type almost always develops before the age of 40 years and has a more severe course. It is inherited as an autosomal dominant Mendelian trait with high penetrance [8].

Primary congenital glaucoma (PCG) is an autosomal recessive disorder caused by a developmental abnormality of the anterior chamber angle and manifests in the neonatal or infantile period. It is characterized by increased IOP, corneal oedema, enlargement of the globe (buphthalmos), epiphora, photophobia and blepharospasm [9]. The incidence of the disease is estimated to be 1:2,000 in the Middle East and 1:10,000 in the Western countries [10]. It has been noted that there is an unequal sex distribution among affected individuals and a lower-than-expected number of affected siblings in familial cases [10–12]. These observations imply that PCG is a genetically heterogeneous disorder.

Glaucoma pathogenesis is multifactorial with significant genetic and environmental contributions. At least 20% of all glaucoma cases have a genetic basis [13], although a recent report suggests that this is an underestimate [14]. Several genes are likely to contribute considerably to the phenotype and will be described in detail in the following.

MYOC

The myocilin gene *MYOC* (MIM601652), also known as trabecular meshwork-inducible glucocorticoid response (*TIGR*) gene, was identified 1997 [15]. It is located on chromosome 1q23–24 and encodes a 504-amino-acid glycoprotein. Mutations in *MYOC* are associated with autosomal dominant JOAG and with POAG. Disease-related mutations have been reported in 2–4% of POAG patients [15] and in up to 33% of JOAG patients [16]. The most common *MYOC* mutation is the heterozygous Q368X mutation which is found in 1.6% of POAG patients [17]. *MYOC* is expressed in almost every ocular tissue, including the optic nerve [18]. It has significant homology with myosin in the amino-terminal region and contains a leucine zipper-like motif similar to that seen in cytoskeletal proteins in the myosin-homology domain [19]. The vast majority of mutations is localized to the terminal third exon which encodes a 250-amino-acid carboxy-terminal domain with homology to olfactomedin [16, 20].

Despite considerable research effort, the function of MYOC remains unknown. Interestingly, recent studies have shown that MYOC +/− and −/− mutant mice have no pathological phenotype. Fertility, viability, IOP, histology and morphology of ocular structures were indistinguishable from wild-type mice [21]. Therefore, it has been suggested that haploinsufficiency is not a critical mechanism for POAG in individuals with mutations in *MYOC* and that disease-related mutations in humans are likely gain-of-function mutations. This hypothesis is confirmed by studies that suggest that POAG is found only in heterozygous patients with one wild-type copy and one mutant copy of *MYOC* [22].

The authors investigated a large French-Canadian family with 622 individuals, of which 83 manifested either JOAG or POAG. Glaucoma developed in 3 heterozygous siblings harbouring a missense mutation at codon 423, whereas 4 siblings that were homozygous for this mutation were asymptomatic for the disease. Therefore, the *MYOC*-associated mutation causes an autosomal dominant but heterozygous-specific phenotype. Morissette et al. [22] suggest that mutant and wild-type forms of the MYOC protein form functionally altered hetero-multimers. The accumulation of these in the cytoplasm or extracellular matrix could impede normal outflow of the aqueous humour and thereby lead to an elevated IOP. Optic neuropathies independent of elevation of IOP could occur by an accumulation of hetero-multimers in retinal ganglion cells or optic nerve fibres. The formation of hetero-multimers has also been suggested by Jacobson et al. [23]. They examined the expression of normal and mutant *MYOC* in patients with and without *MYOC*-associated glaucoma. Very little or no expression of *MYOC* could be detected in cells expressing mutant forms of *MYOC*. The authors suggest that the association of mutant with wild-type myocilin prevents or reduces the secretion of the protein. Recent studies have shown the association of myocilin with the microfibrillar architecture in sheath-derived plaques, a structure where pathologic changes have been documented to occur in eyes of patients with POAG [24]. The elevation of IOP could then induct expression of MYOC in perfused anterior segments [25]. Borrás et al. [25] suggested that this induction pattern might indicate a stress-related rather than a possible homeostatic role for MYOC.

It has been shown that infusion of recombinant MYOC in the anterior chamber of human eyes in organ culture increased outflow resistance and IOP [26]. Therefore it can be assumed that an altered expression of MYOC might also predispose to glaucoma by influencing the uveoscleral outflow. Taken together, these results suggest that mutant MYOC alters the secretion of normal MYOC or other secreted proteins necessary to maintain the integrity of the trabecular meshwork and its extracellular matrix.

CYP1B1

CYP1B1 (MIM601771) is located on chromosome 2p21 and encodes a 543-amino-acid dioxin-inducible member of the cytochrome P_{450} gene superfamily, subfamily I. Mutations in this gene were found in 20–30% of patients with different ethnological backgrounds [27, 28] and up to 85% in consanguineous populations [29, 30]. *CYP1B1* is expressed in different ocular tissues, especially in the iris, trabecular meshwork and ciliary body [31]. It encodes a mono-oxygenase that is capable of metabolizing various endogenous and exogenous substrates, including steroids [32] and retinoids [33]. Identification of *CYP1B1* as the gene affected in PCG is the first example for mutations in a member of the cytochrome P_{450} superfamily resulting in a primary developmental defect [34]. As drug-metabolizing enzymes control the steady-state levels of small bio-organic oxygenated molecules that act as ligands in receptor-mediated signal transduction pathways, it is possible that CYP1B1 plays a role in the metabolism of a yet unknown molecule in eye development. It has been reported that a cytochrome P_{450}-dependent arachidonate metabolite that inhibits Na^+,K^+-ATPase in the cornea was implicated in regulating corneal transparency and aqueous humour secretion [35]. These findings are consistent with the two major diagnostic criteria for primary congenital glaucoma: elevated IOP and clouding of the cornea.

Mutant forms of CYP1B1 may differ in their molecular and enzymatic properties as different combinations of *CYP1B1* mutations lead to phenotypic variability: A subject with Peters' anomaly was found to be a compound heterozygote for two *CYP1B1* mutations [36]; Stoilov et al. [37] described 2 individuals with PCG being compound heterozygotes. These observations suggest that the presence of a particular mutation may not be sufficient to determine the exact phenotypic outcome. In fact, the functional analysis of two different mutant forms of CYP1B1 has shown that they differ in stability and enzymatic activity [38]. Therefore, compound heterozygous individuals for such mutations may exhibit complex biochemical phenotypes, and this observation could account for the phenotypic heterogeneity of *CYP1B1*-associated PCG [37].

Recent studies suggest that CYP1B1 may act as a modifier of MYOC expression [39] although no interaction studies between these two proteins have been performed yet. Vincent [39] showed that individuals with mutations in both genes developed glaucoma much earlier in life than did family members with mutations in *MYOC* alone. They also propose that congenital and juvenile glaucomas are allelic variants.

OPTN

OPTN (optineurin) is the only gene identified so far besides *MYOC* that is associated with adult-onset POAG. The gene was previously identified as *FIP-2* [40]. It is located on chromosome 10p14–p15 and encodes a 577-amino-acid protein that shows no homology to any other known protein but interactions with different proteins like Huntingtin [41], transcription factor IIIA [42] and RAB8 [43] have been documented. Rezaie et al. [44] studied 54 families with autosomal dominant adult-onset glaucoma with at least one member having NTG. They suggested that mutations in *OPTN* may be responsible for 16.7% of all hereditary forms of NTG. Expression of *OPTN* transcripts has previously been reported in heart, brain, placenta, liver, skeletal muscle, kidney and pancreas [40]. Rezaie et al. [44] could also show expression in trabecular meshwork, non-pigmented ciliary epithelium, retina, brain, adrenal cortex, liver, foetus, lymphocytes and fibroblasts. They could also demonstrate that less optineurin is produced in cultured dermal fibroblasts derived from a patient with an E50K missense mutation in *OPTN* than in fibroblasts from a healthy control. This suggests that haploinsufficiency could be the underlying cause of glaucoma. Although it is not known how OPTN is involved in the pathogenesis of glaucoma, it has been shown that it is capable of blocking the protective effect of E3–14.7K on tumour necrosis factor α (TNF-α)-mediated cell killing [40]. That means OPTN can shift the equilibrium towards induction of apoptosis. As TNF-α can markedly increase the severity of damage in optic nerve heads of glaucoma patients [45, 46], it is speculated that OPTN is operating through the TNF-α pathway, playing a neuroprotective role in the eye and optic nerve, but produces optic neuropathy and visual field loss when defective [44].

Loci Linked to Developmental Forms of Glaucoma

Several loci have been investigated in association with developmental forms of glaucoma. Mutations in two genes coding for transcription factors cause a spectrum of glaucoma phenotypes:

PITX2 is located on chromosome 4q25–27 (MIM601542) and encodes a pair-like homeobox transcription factor. It is expressed in developing eye, tooth, umbilicus and pituitary gland [47]. Mutations in *PITX2* are associated with a risk factor for developing glaucoma with developmental anomalies of the anterior segment [48]. Functional assays of mutant PITX2 protein confirm a mutation-specific decrease in DNA binding and altered transactivation properties [49, 50].

FOXC1 is located on chromosome 6p25. It encodes a forkhead transcription factor and was found to be mutated in patients with anterior segment defects. The location of *FOXC1* coincides with glaucoma linked to 6p25 and makes *FOXC1* a prime candidate gene to be investigated [51].

Identification of Susceptibility Factors

As the variability of *MYOC*-associated glaucoma is significant (age of onset, severity, rate of progression, IOP) [17], it is likely that this variability is influenced by factors not yet identified, some of which are probably genetic. *OPA1*, the gene responsible for autosomal dominant optic atrophy, is assumed to be associated with NTG, as a single nucleotide polymorphism (SNP) was found to be strongly associated with the occurrence of the disease [52]. It was also shown that an apolipoprotein E-promoter SNP affects the phenotype of POAG and demonstrates interaction with the myocilin gene [53]. Wiggs et al. [54] used a two-stage genome scan to identify the genomic locations of POAG susceptibility genes. The results of this study have revealed several interesting loci including regions on chromosomes 2, 14, 17 and 19. Further studies will show if genes besides *MYOC* can be identified that contribute to a susceptibility of POAG.

Conclusions and Future Prospects

Early diagnosis is most important for a successful treatment of glaucoma. Therefore, defining the genetic basis of hereditary forms of glaucoma is an important step towards a presymptomatic screening of people at risk.

In the last decade, several chromosomal loci and genes related to glaucoma have been identified, and pedigree-specific highly-penetrant mutations were observed. However, most of these findings relate to juvenile-onset glaucoma or to congenital glaucoma. Today, the *MYOC* gene provides the sole genetic basis for studying the molecular mechanisms that underlie IOP elevation in POAG which is the most frequent type of glaucoma. Several still-to-be-determined genes are likely to play substantial roles in IOP elevation and visual field loss. As a substantial number of glaucoma cases is not associated with an elevated IOP (NTG), genes influencing the retinal ganglion cell layer and the optic nerve head are of particular interest.

In the past 10 years, research on glaucoma has mainly focused on defining susceptibility loci by linkage analyses, determining frequencies in different ethnological populations and studying gene expression by immunohistochemistry. Until recently only few functional studies have been published. Ongoing progress in scientific research however (2-D gel electrophoresis, two hybrid systems, animal models) will facilitate getting insight into protein function and protein interaction.

References

1 Quigley HA: Number of people with glaucoma worldwide. Br J Ophthalmol 1996;80:389–393.
2 Foster PJ, Buhrmann R, Quigley HA, Johnson GJ: The definition and classification of glaucoma in prevalence surveys. Br J Ophthalmol 2002;86:238–242.
3 Tielsch JM, Sommer A, Katz J, Royall RM, Quigley HA, Javitt J: Racial variation in the prevalence of primary open-angle glaucoma. The Baltimore Eye Survey. JAMA 1991;266:369–374.
4 Shiose Y, Kitazawa Y, Tsukahara S, Akamutsu T, Mizokami K, Futa R, Katsushima H: Epidemiology of glaucoma in Japan – A nationwide glaucoma survey. Jpn J Ophthalmol 1991;35: 133–135.
5 Charliat G, Jolly D, Blanchard F: Genetic risk factor in primary open-angle glaucoma: A case control study. Ophthalmic Epidemiol 1994;3:131–138.

6 Wilson MR, Hertzmark E, Walker AM, Childs-Shaw K, Epstein DL: A case-control study of risk factors in open angle glaucoma. Arch Ophthalmol 1987;105:1066–1071.
7 Klein BE, Klein R, Sponsel WE, Franke T, Cantor LB, Martone J, Menage MJ: Prevalence of glaucoma. The Beaver Dam Eye Study. Ophthalmology 1992;99:1499–1504.
8 Sheffield VC, Stone EM, Alward WL, Drack AV, Johnson AT, Streb LM, Nichols BE: Genetic linkage of familial open angle glaucoma to chromosome 1q21–q31. Nat Genet 1993;4:47–50.
9 De Luise V, Anderson DR: Primary infantile glaucoma (congenital glaucoma). Surv Ophthalmol 1983;28:1–19.
10 Jay MR, Rice NSC: Genetic implications of congenital glaucoma. Metab Ophthalmol 1978;2:257–258.
11 Francois J: Congenital glaucoma and its inheritance. Ophthalmologica 1972;181:61–73.
12 Gencik A, Gencikova A, Gerinec A: Genetic heterogeneity of congenital glaucoma. Clin Genet 1980;17:241–248.
13 Wolfs RC, Klaver CC, Ramrattan RS, van Duijn CM, Hofman A, de Jong PT: Genetic risk of primary open-angle glaucoma. Arch Ophthalmol 1998;116:1640–1645.
14 McNaught AI, Allen JG, Healey DL, McCartney PJ, Coote MA, Wong TL, Craig JE: Accuracy and implications of a reported family history of glaucoma: Experience from the Glaucoma Inheritance Study in Tasmania. Arch Ophthalmol 2000;118:900–904.
15 Stone EM, Fingert JH, Alward W: Identification of a gene that causes primary open-angle glaucoma. Science 1997;275:668–670.
16 Shimizu S, Lichter PR, Johnson AT, Zhou Z, Higashi M, Gottfredsdottir M, Othman M, Moroi SE, Rozsa FW, Schertzer RM, Clarke MS, Schwartz AL, Downs CA, Vollrath D, Richards JE: Age-dependent prevalence of mutations at the *GLC1A* locus in primary open-angle glaucoma. Am J Ophthalmol 2000;130:165–177.
17 Fingert JH, Héon E, Liebmann JM, Yamamoto T, Craig JE, Rait J: Analysis of myocilin mutations in 1703 glaucoma patients from five different populations. Hum Mol Genet 1998;8:899–905.
18 Karali A, Russell P, Stefani FH, Tamm ER: Localization of myocilin/trabecular meshwork-inducible glucocorticoid response protein in the human eye. Invest Ophthalmol Vis Sci 2000;41:729–740.
19 Kubota R, Noda S, Wang Y: A novel myosin-like protein (myocilin) expressed in the connecting cilium of the photoreceptor: Molecular cloning, tissue expression and chromosomal mapping. Genomics 1997;41:360–369
20 Adam MF, Belmouden A, Binisti P, Brezin AP, Valtot F, Bechetoilee A, Dascotte JC, Copin B, Gomez L, Chaventre A, Bach JF, Garchon HJ: Recurrent mutations in a single exon encoding the evolutionarily conserved olfactomedin-homology domain of TIGR in familial open-angle glaucoma. Hum Mol Genet 1997;6:2091–2097.
21 Kim BS, Savinova OV, Reedy MV, Martin J, Lun Y, Gan L, Smith RS, Tomarev SI, John SW, Johnson RL: Targeted disruption of the myocilin gene (MYOC) suggests that human glaucoma-causing mutations are gain of function. Mol Cell Biol 2001;21:7707–7713.
22 Morissette J, Clepet C, Moison S, Dubois S, Winstall E, Vermeeren D, Nguyen TD, Polansky JR, Cote G, Anctil JL, Amyot M, Plante M, Falardeau P, Raymond V: Homozygotes carrying an autosomal dominat TIGR mutation do not manifest glaucoma. Nat Genet 1998;19:319–321.
23 Jacobson N, Andrews M, Shepard AR, Nishimura D, Searby C, Fingert JH, Hageman G, Mullins R, Davidson BL, Kwon YH, Alward WL, Stone EM, Clark AF, Sheffield VC: Non-secretion of mutant proteins of the glaucoma gene myocilin in cultured trabecular meshwork cells and in aqueous humour. Hum Mol Genet 2001;10:117–125.
24 Ueda J, Wentz-Hunter K, Yue BY: Distribution of myocilin and extracellular matrix components in the juxtacanalicular tissue of human eyes. Invest Ophthalmol Vis Sci 2002;43:1068–1076.
25 Borrás T, Rowlette LLS, Tamm ER, Gottanka J, Epstein DL: Effects of elevated intraocular pressure on outflow facility and TIGR/MYOC expression in perfused human anterior segments. Invest Ophthalmol Vis Sci 2002;43:33–40.
26 Fautsch MP, Bahler CK, Jewison DJ, Johnson DH: Recombinant TIGR/MYOC increases outflow resistance in the human anterior segment. Invest Ophthalmol Vis Sci 2000;41:4163–4168.
27 Héon E, Martin N, Billingsley G, Williams-Lyn D, Sutherland J, Levin A: Molecular characterization of congenital glaucoma in the Greater Toronto Area. Invest Ophthalmol Vis Sci Suppl 2000;41:527.

28 Kakiuchi-Matsumoto T, Isashiki Y, Ohba N, Kimura K, Sonoda S, Unoki K: Cytochrome P_{450}1B1 gene mutations in Japanese patients with primary congenital glaucoma. Am J Ophthalmol 2001;131:345–350.

29 Bejjani BA, Lewis RA, Tomey KF, Anderson KL, Dueker DK, Jabak M, Astle WF, Otterud B, Leppert M, Lupski JR: Mutations in CYP1B1, the gene for cytochrome P_{450}B1, are the predominant cause of primary congenital glaucoma in Saudi Arabia. Am J Hum Genet 1998;62:325–333.

30 Plášilová M, Stoilov I, Sarfarazi M, Kádasi L, Feráková E, Ferák V: Identification of a single ancestral *CYP1B1* mutation in Slovak gypsies (Roms) affected with primary congenital glaucoma. J Med Genet 1999;36:290–294.

31 Sutter TR, Tang YM, Hayes CL, Wo YYP, Jabs EW, Li X, Yin H, Cody CW, Greenlee WF: Complete cDNA sequence of a human dioxin-inducible mRNA identifies a new gene subfamily of cytochrome P_{450} that maps to chromosome 2. J Biol Chem 1994;269:13092–13099.

32 Hayes CL, Spink DC, Spink BC, Cao JQ, Walker NJ, Sutter TR: 17β-Estradiol hydroxylation catalyzed by cytochrome P_{450}B1. Proc Natl Acad Sci USA 1997;93:9776–9781.

33 Zhang QY, Dunbar D, Kaminsky L: Human cytochrome P_{450} metabolism of retinals to retinoic acids. Drug Metab Dispos 2000;28:292–297.

34 Stoilov I, Akarsu AN, Sarfarazi M: Identification of three truncating mutations in cytochrome P_{450}1B1 (*CYP1B1*) as the principal cause of primary congenital glaucoma (buphthalmos) in families linked to the *GLC3A* locus on chromosome 2p21. Hum Mol Genet 1997;6:641–647.

35 Schwartzman ML, Balazy M, Masferrer J, Abraham NG, McGiff JC, Murphy RC: 12(*R*)-Hydroxyeicosatetraenoic acid: A cytochrome P_{450}-dependent arachidonate metabolite that inhibits Na^+,K^+-ATPase in the cornea. Proc Natl Acad Sci USA 1987;84:8125–8129.

36 Vincent A, Billingsley G, Priston M: Phenotypic heterogeneity of CYP1B1: Mutations in a patient with Peters' anomaly. J Med Genet 2001;38:324–326.

37 Stoilov IR, Costa VP, Vasconcellos JPC, Melo MB, Betinjane AJ, Carani JCE, Oltrogge EV, Sarfarazi M: Molecular genetics of primary congenital glaucoma in brazil. Invest Ophthalmol Vis Sci 2002;43:1820–1827.

38 Jansson I, Stoilov I, Sarfarazi M, Schenkman JB: Effect of two mutations of human CYP1B1, G61E and R469W on stability and endogenous steroid substrate metabolism. Pharmacogenetics 2001;11:793–801.

39 Vincent AL, Billingsley G, Buys Y, Levin AV, Priston M, Trope G, Williams-Lyn D, Héon E: Digenic inheritance of early-onset glaucoma: CYP1B1, a potential modifier gene. Am J Hum Genet 2002;70:448–460.

40 Li Y, Kang J, Horwitz, MS: Interaction of an adenovirus E3 14.7-kilodalton protein with a novel tumor necrosis factor-α-inducible cellular protein containing leucine zipper domains. Mol Cell Biol 1998;18:1601–1610.

41 Faber PW, Barnes GT, Srinidhi J, Chen J, Gusella JF, MacDonald ME: Huntingtin interacts with a family of WW domain proteins. Hum Mol Genet 1998;7:1463–1474.

42 Moreland RJ, Dresser ME, Rodgers JS, Roe BA, Conaway RC, Conaway JS, Hanas JS: Identification of a transcription factor IIIA-interacting protein. Nucleic Acids Res 2000;28:1986–1993.

43 Hattula K, Peranen J: FIP-2, a coiled-coil protein, links Huntingtin to Rab8 and modulates cellular morphogenesis. Curr Biol 2000;10:1603–1606.

44 Rezaie T, Child A, Hitchings R, Brice G, Miller L, Coca-Prados M, Héon E, Krupin T, Ritch R, Kreutzer D, Crick RP, Sarfarazi M: Adult-onset primary open-angle glaucoma caused by mutations in optineurin. Science 2002;295:1077–1079.

45 Yuan L, Neufeld AH: Tumor necrosis factor-α: A potentially neurodestructive cytokine produced by glia in the human glaucomatous optic nerve head. Glia 2000;32:42–50.

46 Tezel G, Wax MB: Increased production of tumor necrosis factor-α by glial cells exposed to simulated ischemia or elevated hydrostatic pressure induces apoptosis in cocultured retinal ganglion cells. J Neurosci 2000;23:8693–8700.

47 Semina EV, Reiter R, Leysens NJ, Alward WL, Small KW, Datson NA, Siegel-Bartelt J, Bierke-Nelson D, Bitoun P, Zabel BU, Carey JC, Murray JC: Cloning and characterization of a novel bicoid-related homeobox transcription factor gene, RIEG, involved in Rieger syndrome. Nat Genet 1996;14:392–399.

48 Kulak SC, Kozlowski K, Semina EV, Pearce WG, Walter MA: Mutation in the RIEG1 gene in patients with iridogoniodysgenesis syndrome. Hum Mol Genet 1998;7:1113–1117.
49 Kozlowski K, Walter MA: Variation in residual PITX2 activity underlies the phenotypic spectrum of anterior segment developmental disorders. Hum Mol Genet 2000;9:2131–2139.
50 Priston M, Kozlowski K, Gill D, Letwin K, Buys Y, Levin AV, Walter MA, Héon E: Functional analyses of two newly identified PITX2 mutants reveal a novel molecular mechanism for Axenfeld-Rieger syndrome. Hum Mol Genet 2001;10:1631–1638.
51 Lehmann OJ, Ebenezer ND, Ekong R, Ocaka L, Mungall AJ, Fraser S, McGill JI, Hitchings RA, Khaw PT, Sowden JC, Povey S, Walter MA, Bhattacharya SS, Jordan T: Ocular developmental abnormalities and glaucoma associated with interstitial 6p25 duplications and deletions. Invest Ophthalmol Vis Sci 2002;43:1843–1849.
52 Aung T, Ocaka L, Ebenezer ND, Morris AG, Krawczak M, Thiselton DL, Alexander C, Votruba M, Brice G, Child AH, Francis PJ, Hitchings RA, Lehmann OJ, Bhattacharya SS: A major marker for normal tension glaucoma: Association with polymorphisms in the OPA1 gene. Hum Genet 2002;110:52–56.
53 Copin B, Brézin AP, Valtot F, Dascotte JC, Béchetoille A, Garchon HJ: Apolipoprotein E-promoter single-nucleotide polymorphisms affect the phenotype of primary open-angle glaucoma and demonstrate interaction with the myocilin gene. Am J Hum Genet 2002;70:1575–1581.
54 Wiggs JL, Allingham RR, Hossain A, Kern J, Auguste J, DelBono EA, Broomer B, Graham FL, Hauser M, Pericak-Vance M, Haines JL: Genome-wide scan for adult-onset primary open angle glaucoma. Hum Mol Genet 2000;9:1109–1117.

Nicole Weißschuh
Molekulargenetisches Labor, Universitäts-Augenklinik, Tübingen
Auf der Morgenstelle 15, D–72076 Tübingen (Germany)
Tel. +49 7071 2987618, Fax +49 7071 295725, E-Mail nicole.weissschuh@uni-tuebingen.de

Wissinger B, Kohl S, Langenbeck U (eds): Genetics in Ophthalmology.
Dev Ophthalmol. Basel, Karger, 2003, vol 37, pp 94–108

LHON and Other Optic Nerve Atrophies: The Mitochondrial Connection

Neil Howell

MitoKor, San Diego, Calif. and Departments of Radiation Oncology, and Human Biological Chemistry and Genetics, The University of Texas Medical Branch, Galveston, Tex., USA

Abstract

The clinical, biochemical and genetic features of Leber's hereditary optic neuropathy (LHON) are reviewed. The etiology of LHON is complex, but the primary risk factor is a mutation in one of the seven mitochondrial genes that encode subunits of respiratory chain complex I. The pathogenesis of LHON is not yet understood, but one plausible model is that increased or altered mitochondrial ROS production renders the retinal ganglion cells vulnerable to apoptotic cell death. In addition to LHON, there are a large number of other optic nerve degenerative disorders including autosomal dominant optic atrophy, the toxic/nutritional optic neuropathies and glaucoma. A review of the recent scientific literature suggests that these disorders also involve mitochondrial dysfunction or altered mitochondrial signaling pathways in their pathogenesis. This mitochondrial link provides new avenues of experimental investigation to these major causes of loss of vision.

Introduction

It is estimated that more than 75 human diseases involve, in some part of the pathogenesis, mitochondrial dysfunction. Mitochondria are best known as the predominant source of cellular energy production, but it is also becoming clear that they play a broader role in cellular metabolism, signaling, structure and genetics. From the standpoint of pathology, mitochondria play a key role in cell death pathways, both apoptotic and necrotic. In this chapter, the main features of Leber's hereditary optic neuropathy (LHON), an inherited optic atrophy in which a mitochondrial pathophysiology is certain, are briefly

Table 1. Summary of the effects of the major LHON mutations

	LHON mutation		
	3460	11778	14484
Mitochondrial gene	ND1	ND4	ND6
Amino acid position and change	52/A→T	340/R→H	64/M→V
Prevalence, %[a]	10–15	60–70	15–20
Male predominance, %[b]	~70	70–85	70–85
Mean age of onset, years	~29	~28	25–27
Recovery	22–29	2–4	36–50
Time to nadir, months	2–3	2–4	2–4
Visual nadir – 6/60 or worse, %	67–75	70–95	29–47

[a]The prevalence is expressed as the approximate percentage of LHON pedigrees of European descent that carry the particular LHON mutation. These figures do not include 'LHON plus' families, but these represent only a small fraction (<5%) of typical LHON families.

[b]Male predominance is indicated as the percentage of all affected individuals (i.e., males plus females). For this parameter and the others that follow, the ranges provide an indication of the values obtained in different studies. Data were extracted only from recent studies [reviewed in 2–5], and they do not necessarily reflect historical values. Finally, percentages for recovery and visual nadir refer to the total number of affected individuals, not to the total number of at-risk and affected family members.

reviewed. In addition to LHON, it is now emerging that other optic nerve degenerative disorders also involve mitochondrial abnormalities. It is this 'mitochondrial connection' that is the focus of this review.

Leber's Hereditary Optic Neuropathy

LHON (MIM 535000) was first described in detail by Theodor Leber in 1871 [1]. Typically, LHON is an acute or subacute bilateral loss of central vision, usually painless, that manifests in the second to fourth decades of life (table 1). The only abnormality noted at the presymptomatic stage is a peripapillary microangiopathy in a substantial proportion of family members that subsequently resolves. In the acute phase, there is onset of the vision abnormalities (the first sign is usually a blurring of vision), pseudoedema of the nerve fiber layer, and hyperemia of the optic disk. At the atrophic stage, the vision abnormalities have reached their nadir, the disk flattens and becomes pale, and the peripapillary nerve fiber layer thins or disappears, especially in the region near the papillomacular bundle. Detailed reviews of the pathophysiology, epidemiology and genetics of LHON are available [2–6]. Beginning with the affected families of Leber [1], it was recognized that the risk of developing the optic neuropathy showed *maternal* transmission. It is now understood that the primary etiological event or factor in LHON is a mutation in the mitochondrial DNA (mtDNA), which is strictly maternally inherited. In fact, LHON appears to be the most prevalent mitochondrial genetic disorder [7].

The optic neuropathy in LHON is due to a loss of retinal ganglion cells and a degeneration of the optic nerve, almost certainly through a mitochondrial apoptotic pathway, rather than a necrotic one [4, 5]. LHON has a very similar clinical presentation to the autosomally inherited optic neuropathies and to the toxic-nutritional optic neuropathies (see below). In particular, LHON involves a preferential loss of the P-ganglion cells (those with smaller cell diameters) that subserve central vision [6, 8]. Careful analysis of the pathology by Carelli et al. [6] has led them to conclude that, during the atrophic stage, there is a slow but ongoing loss of retinal ganglion cells that can continue for decades. They have also called attention to abnormalities of the myelin sheath and to the ultrastructural evidence for impaired axonal transport. There is a remarkable concentration of mitochondria in the optic nerve head, which is the region where the nerve fibers remain unmyelinated and transverse the lamina cribosa [4, 5]. This region constitutes an 'energetic chokepoint' that is particularly sensitive to disruption of mitochondrial function (see further discussion below).

In the vast majority of LHON patients, the optic neuropathy is the sole clinical abnormality, although there appear to be increased frequencies of other neurological deficits, most notably an MS-like syndrome, among LHON family members. It has recently been estimated that 5% of LHON pedigrees include at least one family member with MS [9]. Although one suspects that careful examination would reveal a higher incidence of relatively subtle neurological and ophthalmological abnormalities, the point remains that the pathology is remarkably limited to the optic neuropathy. In contrast to this typical presentation, there are a small proportion of 'LHON plus' families in which there are severe abnormalities, such as dystonia or encephalopathy, that overshadow the optic neuropathy.

LHON has an incomplete penetrance and most family members remain clinically unaffected throughout life. This key feature indicates that the LHON mtDNA mutation is necessary, but not sufficient, for manifestation of the optic neuropathy. Rather, the sudden onset of the vision abnormalities suggests that secondary factors exacerbate the already compromised mitochondrial metabolism beyond a crisis point where the optic nerve can no longer function, and where the acute phase is 'triggered' [4, 5]. As will be developed in subsequent sections, the concept of a complex or 'multi-hit' etiology, one component of which involves mitochondrial metabolism or signaling, is one that extends to other degenerative optic nerve disorders. As a side issue, it should be noted that we use the term 'penetrance' to be consistent with previous usage. The terms 'penetrance' and 'expressivity' both have elements that are appropriate and others that are not.

In contrast to the other optic atrophies discussed here, males are affected approximately 4–5 times more often than females in LHON families (table 1). The obvious explanation for this disparity is an X-linked modifier locus, but the accumulating experimental data argue against this simple explanation (see Wittig et al. [10] for the most recent results, and references therein).

Since the identification of the first LHON mtDNA mutation in 1988 [11], hundreds of LHON patients and family members have been analyzed to identify the pathogenic mtDNA mutations. There is broad agreement that approximately 95% of classic LHON cases in populations of European descent result from one of three mtDNA mutations genes that encode subunits of respiratory chain complex I [e.g., 12]: (a) a G:A transition at nucleotide 3460 that results in the substitution of THR for ALA at amino acid position 52 of the ND1 gene; (b) a G:A transition at nucleotide 11778 which changes ND4/ARG340 to HIS; or (c) a T:C transition at nucleotide 14484 which changes ND6/MET64 to VAL (table 1). A number of other mtDNA mutations that cause LHON have been identified, and these also occur in mitochondrial genes that encode complex I subunits [13, 14]. In ~15% of LHON patients, the mtDNA mutation is heteroplasmic (that is, both wild-type and mutant copies of the gene are present in an individual). As would be expected, the risk of developing the optic neuropathy is related to mutation load, and males with a load of the 11778 LHON mutation in blood of less than 60% have a low (but not zero) risk of vision loss [15].

There are only minor differences among these three LHON mutations in terms of the optic neuropathy (table 1), although there is a clear disparity in spontaneous recovery of vision after the acute phase [e.g., 2, 16]. Recovery of vision is very rare in 11778 LHON patients, but it is relatively frequent in 14484 LHON patients (~50% if vision is lost before the age of 30 years). It is interesting to note that the extent of vision loss is apparently less severe, as well, in 14484 LHON patients. However, this trend is probably related to the high rate of recovery, with vision loss not being accurately assessed in a substantial proportion of 14484 LHON patients until after recovery is already underway. There is a recent report [17] in which the measured frequency of functional recovery is higher if one analyzes visual fields (static perimetry) and critical flicker frequency, in addition to visual acuity. These results suggest that the frequency of recovery may have been underestimated.

Recovery of vision is providing important – but not as yet understood – 'clues' about the underlying biochemical and neurochemical abnormalities. It means that, in a substantial proportion of LHON patients, there can be a prolonged period of time in which retinal ganglion cells lose function but do not die. Carelli et al. [6] have raised an interesting possibility. On the basis of their extensive histopathological studies, they emphasize the importance of the optic nerve demyelination in LHON, and they suggest that vision recovery may be due to remyelination of optic nerve axons [for a different view, see 5]. They have also raised the possibility that the MS-like condition in some LHON family members may be due to a more 'global' demyelination beyond the retrobulbar optic nerve.

The 'LHON plus' families display a wider array – and greater severity – of clinical abnormalities, but these complex disorders are less well understood and occur in a relatively small number of matrilineal pedigrees. For example, Shoffner et al. [18] have shown that a G:A mutation at nucleotide 14459 (ND6/ALA72 changed to VAL) causes LHON plus dystonia in three unrelated families. However, the same mtDNA mutation occurs in two unrelated Australian families who are affected with Leigh syndrome (subacute necrotizing encephalomyelopathy), but who have no history of optic neuropathy or dystonia [19]. De Vries et al. [20] reported that a large family in which LHON was associated with hereditary spastic dystonia carried both a heteroplasmic mutation at nucleotide 11696 (ND4/VAL312 changed to ILE) and a homoplasmic mutation at nucleotide 14596 (ND6/MET26 changed to ILE). Either mutation, or both, may be the primary pathogenic factor. There is a relatively large Australian LHON pedigree (designated QLD1) in which the optic neuropathy is accompanied by a variety of severe neurological abnormalities, including a fatal infantile encephalopathy. The optic neuropathy appears to be caused by the 14484 LHON mutation, whereas the neurological abnormalities are caused by a mutation at nucleotide 4160 that changes the ND1/LEU285 amino acid residue to PRO [21]. Finally, mutations at nucleotides 13513 (ND5/ASP393 changed to ASN) and 13514 (ND5/ASP393 changed to GLY) are associated with a LHON/MELAS overlap syndrome [22, 23]. The neurological abnormalities are lumped under the term 'MELAS' (mitochondrial encephalopathy with lactic acidosis and stroke-like episodes), but they are quite heterogeneous and include bilateral hearing loss, pyramidal signs, memory loss, and muscle atrophy in addition to the characteristic stroke-like episodes.

There are now four 'two mutation' LHON families that carry both the 11778 and 14484 mutations. In one family, 2 sisters have had 5 children, all of whom died from a fatal infantile encephalomyopathy [24]. However, the general finding is that 'two mutation' LHON family members are not affected more often or more severely than individuals who carry a single LHON mutation.

In addition to the mtDNA mutations with a primary pathogenic role in LHON, other mtDNA polymorphisms may comprise one type of secondary etiologic risk factor. Approximately 75% of the mtDNAs from 14484 LHON pedigrees of European descent belong to haplogroup J (a haplogroup is a set of sequences of common evolutionary descent that occur within a single major ethnic group), although only ~10% of mtDNAs from the general European population belong to this haplogroup. There also appears to be a weak haplogroup J association for the 11778 LHON mutation, but not for the 3460 LHON mutation. It has been proposed that this haplogroup clustering reflects a higher penetrance of the 14484 and 11778 LHON mutations when they are 'embedded' within a haplogroup J background [25, 26]. According to this explanation, one or more polymorphisms within haplogroup J mtDNAs have a phenotype that increases penetrance of the 14484 LHON mutation. Because penetrance thus varies according to the mtDNA background, there is an incomplete ascertainment bias. Families who carry a LHON mutation, but who lack affected members, will not come to the attention of clinicians and researchers. However, penetrance in LHON pedigrees is affected by multiple, poorly-defined secondary factors, and conclusive experimental support for this hypothesis has been difficult to obtain.

The fact that the primary cause of LHON is a mutation in a mitochondrial complex I gene might have been expected to yield a predictable relationship between complex I activity and one or more characteristics of the optic neuropathy (e.g., frequency of vision recovery or severity of vision loss). Such an expectation, however, has not been met and 10 years of biochemical studies suggest that the optic neuropathy involves more than a simple complex I catalytic defect [reviewed in 4, 27]. The recent results of Brown et al. [28] agree with the bulk of earlier biochemical studies, and they serve as a good example of the unresolved complexities. These investigators analyzed both lymphoblastoid lines from LHON patients and transmitochondrial cybrid lines into which the 3460, 11778, 14484, or 14459 LHON mutations had been transferred. These mutations rank in the order 14484, 3460, 11778 and 14459 (least severe to most severe) using likelihood of vision recovery and prevalence of associated neurological abnormalities as ranking criteria. In contrast to this order, the 14459 mutation caused no measurable defect in the rate of electron transfer through the entire span of the chain (respiration); the 14484 mutation produced only a mild defect, and the 3460 and 11778 mutations produced slightly greater defects in cybrid lines. In contrast, the 14459 and 3460 mutations produced marked reductions (60–70%) in complex I activity, whereas the 11778 and 14484 mutations caused little, if any, reduction [see also 29]. The 'disconnect' between respiration rates and complex I activities is unexpected, based on what we know about the relationship between the two. The analysis of mitochondrial function in fibroblasts from 3460 LHON patients is also noteworthy [30]. As with other biochemical studies, a marked decrease (~60%) in complex I activity was measured, but the key observation was that there was *no* significant impairment in ATP production. When the same assays were repeated with fibroblasts from a patient with mitochondrial cardiomyopathy, there were defects – as one would predict – in *both* complex I activity and ATP production. These results suggest that biochemical studies, involving cell disruption and isolation of mitochondria, may somehow alter respiratory chain function in cells that carry LHON mutations.

There would be some sense of comfort if the biochemical studies were in better agreement with the reports of in situ assays of mitochondrial respiratory chain activity, but they do not. For example, Lodi et al. [31, 32] used ^{31}P magnetic resonance spectroscopy to assess energy metabolism in skeletal muscle of LHON patients. In contrast to the biochemical studies with isolated mitochondria, they observed that the maximal rate of ATP synthesis was ~30% of normal during the exercise recovery phase in 11778 LHON patients. The equivalent rate was ~50% in patients who harbored the 14484 LHON mutation, an order that is reversed from the biochemical studies. The most striking result was that the 3460 LHON mutation was associated with, at most, a mild defect in muscle energy production. The most recent results [33] confirm that the 3460 mutation does not compromise bioenergetic function in muscle. However, a severe defect in brain bioenergetics was observed. The reason for the marked tissue-specific expression of the mutation phenotype is not understood, but these results will be an important stepping-off point for further investigation of mitochondrial disorders.

Taken together, these results indicate that the vision loss in LHON does not result from a general defect in mitochondrial electron transfer, but from some specific perturbation of complex I. As an illustration of this point, Chinnery et al. [14] carried out secondary structure modeling of the ND6 subunit of complex I. They found that the several LHON mutations analyzed were localized to a specific region of the subunit that involved a transmembrane hydrophobic cleft or pocket and residues in the adjacent extramembrane hydrophilic loops. This is the first evidence of a relationship between a mitochondrial disease and a specific region of a respiratory chain subunit. The question is, however, what is the function of this region and how does its alteration lead to LHON?

In view of the evidence that the optic neuropathy in LHON is not a simple result of decreased complex I activity, other pathogenic pathways have been considered. Both Brown et al. [13] and Sadun et al. [8] have suggested that LHON mutations increase mitochondrial ROS (reactive oxygen species) production, probably through perturbation of the quinone redox reactions in complex I [e.g., 29]. This increased ROS production would lead in turn to oxidative stress, triggering retinal ganglion cell dysfunction and – eventually – apoptotic cell death. Klivenyi et al. [34] have reported that the concentration of plasma α-tocopherol was reduced in 11778 LHON family members, irrespective of whether they were visually affected or unaffected, and they suggested that this result reflected increased oxidative stress. Wong and Cortopassi [35] have shown that LHON cybrid cells are more sensitive to killing by ROS, suggesting that the antioxidant systems of these cells may have been 'overloaded' as some consequence of mitochondrial dysfunction.

Experimental evidence for a ROS-mediated apoptotic pathogenesis has come from two recent studies of Cortopassi and co-workers. In the first study, they showed that osteosarcoma cybrid lines that carried the 11778 or 3460 LHON mutations were more sensitive to Fas-induced apoptosis than control cybrid lines [36]. In the second study, these LHON mutations were introduced into neuronal NT2 cybrid cell lines [37]. They observed that, in undifferentiated LHON NT2 cybrid lines, mitochondrial ROS production was not increased above the levels produced by undifferentiated control cybrid lines. However, a different pattern was obtained when these cybrid lines were differentiated into neuronal cell populations. Under those conditions, the LHON mutations were associated with increased levels of mitochondrial ROS production and the authors suggested that their results might provide an explanation for the focal pathology of LHON (see above). As appealing as is this suggestion, other studies have indicated that retinal ganglion cells are *more* resistant to ROS-induced cell killing than are other retinal cell types [38]. Perhaps LHON mutations produce a specific type of mitochondrial free radical damage that is especially or selectively toxic to retinal ganglion cells. It is fair to say, in summary, that our understanding of the pathway that connects LHON mtDNA mutations to selective death of retinal ganglion cells remains incomplete.

A ROS/oxidative stress mechanism for LHON suggests an explanation for the predominance of affected males relative to females. Females might be less vulnerable to development of the optic neuropathy because of their higher estrogen levels during the 'time window' when risk is highest. There is compelling evidence that estrogen compounds are both neuroprotective and neurotrophic, and – furthermore – that estrogens can act on mitochondria to relieve oxidative stress [39].

Autosomal Dominant Optic Atrophy and Glaucoma

Autosomal dominant optic atrophy (ADOA) typically presents in childhood with slow bilateral vision loss, optic nerve pallor, color vision abnormalities, and centrocecal defects in the visual fields [40, 41]. The presentation of the ophthalmological abnormalities is variable, even among patients of the same age. Despite the different time courses, the ultimate pathology of ADOA is very similar to that of LHON. In fact, the two disorders can be difficult to distinguish clinically [e.g., 42, 43]. It has recently been shown that a large number of ADOA cases are caused by mutations in the *OPA1* gene [43–47; MIM165500]. The protein encoded by the *OPA1* gene is a widely expressed member of the GTP-binding dynamin family and it is particularly abundant in the retina. Most importantly, this protein is localized in the mitochondrion and, on the basis of its possible homology to the yeast *Mgm1/Msp1* proteins, it is probably necessary for the proper maintenance of mitochondrial structural and genetic integrity. There is one report that the *OPA1* mutations alter the structure of the mitochondrial reticulum in monocytes [45], but no changes in leukocyte mitochondria were observed in another [47]. It is obviously important to determine if structural abnormalities are present in the mitochondria of retinal ganglion cells from patients.

There is emerging evidence that there is also a mitochondrial link in glaucoma, an optic neurodegenerative condition that affects millions of patients around the world. Actually, 'links' may be a more appropriate term, because the death of the retinal ganglion cells in glaucoma – as in LHON – proceeds through a mitochondrial apoptotic pathway [reviewed in 4, 5]. The potential of drugs that act on mitochondria and prevent the early steps in the apoptotic cascade for the treatment of glaucoma has been discussed elsewhere [48]. In addition, there appears to be another mitochondrial link to glaucoma, this one occurring at an earlier stage of the pathology. The first piece of evidence is that an association has been found [49] between polymorphisms in the *OPA1* gene, which as described above is involved in ADOA, and normal tension glaucoma (MIM 605290), a major subclass of primary open-angle glaucoma (POAG). On the other hand, no association has been found between the presence of LHON mtDNA mutations and normal tension glaucoma [50].

A particularly important set of studies that link mitochondria to glaucoma involves the role of mutations in the *GLC1* gene and glaucoma. Mutations in the GLC1 gene are a major cause of inherited, juvenile-onset POAG, but mutations have also been detected in a small proportion (~2–4%) of apparently-sporadic, adult-onset POAG [e.g., 51–53]. The *GLC1* gene encodes a protein that has been designated myocilin or trabecular meshwork-inducible glucocorticoid response protein (TIGR; MIM601652). Myocilin is inducible in trabecular meshwork cell cultures by glucocorticoids, but the important point for this discussion is that a substantial proportion of the protein is associated with the mitochondrion [54]. Furthermore, these investigators observed that either the induction of myocilin, or an increased association of myocilin with mitochondria, renders trabecular meshwork cells more susceptible to triggers of apoptotic cell death. The precise mechanism by which *GLC1* mutations cause death of retinal ganglion cells is not yet known, but the mitochondrial link clearly provides a new avenue of investigation.

At the risk of overinterpreting the available reports, there is another mitochondrial connection to glaucoma. Rezaie et al. [55] have obtained evidence that mutations in the *OPTN* gene are the major etiological event in a substantial proportion of familial and sporadic cases of adult-onset normal tension glaucoma (MIM 602432). These investigators also showed that the gene product, optineurin, is localized to the Golgi apparatus. Optineurin interacts with a number of intriguing regulatory proteins [55], and it appears to play a neuroprotective role through its interactions with the TNF-α apoptotic pathway. What is the mitochondrial connection? The TRAP-1 protein (tumor necrosis factor receptor-associated protein 1) is predominantly localized to mitochondria in a number of cell types [56], although its intracellular distribution in the retina is not yet known. A number of apoptotic pathways involve proteins moving into, and out of, the mitochondrion. It is a plausible scenario that optineurin may function normally, at least in part, to 'dampen' apoptosis by blocking TNF-α trafficking to mitochondria [see also 57]. *OPTN* mutations may predispose the retinal ganglion cells to apoptosis, and glaucoma is manifested when other specific risk factors occur (see the discussion of a two-hit model for glaucoma in Wentz-Hunter et al. [54]).

Toxic and Nutritional Optic Neuropathies

It was observed in the 1950s and 1960s that an optic neuropathy with a similar focal pathology to LHON was caused, as a side effect, by systemic treatment of humans with chloramphenicol, an antibacterial agent that also specifically inhibits mitochondrial protein synthesis [reviewed in 4]. These reports are particularly important because they are the first example of a toxic optic neuropathy whose pathogenesis can be traced to mitochondrial respiratory chain dysfunction. An optic neuropathy is also a side effect of treatment with ethambutol, an antimycobacterial drug. Studies of the cytotoxic effects of ethambutol on rat retinal ganglion cells indicated that the drug acts through a glutamate excitotoxic pathway in which mitochondrial calcium levels are increased, and in which there is evidence of decreased mitochondrial respiratory chain function [58]. However, another study [59], while validating toxicity of ethambutol towards retinal ganglion cells, found no evidence for an excitotoxic mechanism in retinal cell cultures, so the mitochondrial link here remains tenuous at this stage.

Carelli et al. [6] have recently pointed to the similar pathologies of a number of acquired optic neuropathies, in addition to those caused by chloramphenicol and by ethambutol, and that of LHON. They, and this is a point of strong agreement on my part [4, 5], emphasize the mitochondrial energetic 'chokepoint' as the link among these different optic neuropathies. The region of the optic nerve most susceptible to the loss of energy production is the papillomacular bundle, particularly in the prelamellar region of the optic nerve head where the nerve fibers are both unmyelinated and sharply bent and where there is a marked concentration of mitochondria [see especially 60]. This is also the region of the optic nerve that is affected earliest and most severely during the acute phase of LHON. Carelli et al. [6] point out that toxins or conditions that compromise energy metabolism are thus likely to affect this 'chokepoint' and, as a result, to produce a similar pathology. They review a number of toxic and nutritional optic neuropathies that support this view.

Conclusions and Future Directions

The point of this review has been to highlight the involvement of mitochondrial pathways in a number of optic nerve degenerative disorders. One obvious future direction is verification and clarification. While a mitochondrial 'connection' among the optic neuropathies is gaining support, there is – as yet at least – no single mitochondrial pathway or mechanism. Instead, there are clear differences in etiology and pathogenesis among these disorders, strongly suggesting multiple pathways. The common feature might be the vulnerability of retinal ganglion cells to oxidative stress or bioenergetic compromise. Future research will continue to answer the questions that we have now, in addition (of course) to raising new ones. At the same time, the emerging role of mitochondria in these optic nerve disorders also provides a new avenue for treatment. At present, there are no broadly effective treatments for any of these disorders. Fortunately, there is an increasing interest in mitochondrial targets for drug development [48, 61], so that we may soon be better placed to treat these devastating disorders.

Acknowledgements

The support of my research on LHON by the Eierman Foundation is gratefully acknowledged. I also acknowledge the contributions of my collaborators Drs. Doug Turnbull and Patrick Chinnery (University of Newcastle) and Dr. David Mackey (Royal Victoria Eye and Ear Hospital, Melbourne).

References

1 Leber T: Über hereditäre und congenital angelegte Sehnervenleiden. Graefes Arch Ophthalmol 1871;17:249–291.

2 Riordan-Eva P, Sanders MD, Govan GG, Sweeney MG, Da Costa J, Harding AE: The clinical features of Leber's hereditary optic neuropathy defined by the presence of a pathogenic mitochondrial DNA mutation. Brain 1995;118:319–338.

3 Kleiner L, Sherman J: Leber's hereditary optic neuropathy: Historical and contemporary considerations. Optom Clin 1996;5:77–112.

4 Howell N: Leber hereditary optic neuropathy: Mitochondrial mutations and degeneration of the optic nerve. Vision Res 1997;37:3495–3507.

5 Howell N: Leber hereditary optic neuropathy: Potential opportunities/potential pitfalls for drug therapy of optic nerve degenerative disorders. Drug Dev Res 1999;46:34–43.

6 Carelli V, Ross-Cisneros FN, Sadun AA: Optic nerve degeneration and mitochondrial dysfunction: Genetic and acquired optic neuropathies. Neurochem Int 2002;40:573–584.

7 Chinnery PF, Johnson MA, Wardell TM, Singh-Kler R, Hayes C, Brown DT, Taylor RW, Bindoff LA, Turnbull DM: The epidemiology of pathogenic mitochondrial DNA mutations. Ann Neurol 2000;48:188–193.

8 Sadun AA, Win PH, Ross-Cisneros FN, Walker SO, Carelli V: Leber's hereditary optic neuropathy differentially affects smaller axons in the optic nerve. Trans Am Ophthalmol Soc 2000;98:223–235.

9 Vanopdenbosch L, Dubois B, D'Hooghe MB, Meire F, Carton H: Mitochondrial mutations of Leber's hereditary optic neuropathy: A risk factor for multiple sclerosis. J Neurol 2000;247:535–543.

10 Wittig I, Augstein P, Brown GK, Fujii T, Rötig A, Rustin P, Munnich A, Seibel P, Thorburn D, Wissinger B, Tamboom K, Metspalu A, Lamantea E, Zeviani M, Wehnert MS: Sequence variations in the *NDUFA1* gene encoding a subunit of complex I of the respiratory chain. J Inherit Metab Dis 2001;24:15–27.

11 Wallace DC, Singh G, Lott MT, Hodge JA, Schurr TG, Lezza AMS, Elsas LJ, Nikoskelainen E: Mitochondrial DNA mutation associated with Leber's hereditary optic neuropathy. Science 1988;242:1427–1430.

12 Mackey DA, Öostra RJ, Rosenberg T, Nikoskelainen E, Bronte-Stewart J, Poulton J, Harding AE, Govan G, Bolhuis PA, Norby S, Bleeker-Wagemakers EM, Savontaus ML, Chan C, Howell N: Primary pathogenic mitochondrial DNA mutations in multigeneration pedigrees with Leber hereditary optic neuropathy. Am J Hum Genet 1996;59:481–485.

13 Brown MD, Zhadanov S, Allen JC, Hosseini S, Newman NJ, Atamonov VV, Mikhailovskaya IE, Sukernik RI, Wallace DC: Novel mtDNA mutations and oxidative phosphorylation dysfunction in Russian LHON families. Hum Genet 2001;109:33–39.

14 Chinnery PF, Brown DT, Andrews RM, Singh-Kler R, Riordan-Eva P, Lindley J, Appelgarth DA, Turnbull DM, Howell N: The mitochondrial *ND6* gene is a hot spot for mutations that cause Leber's hereditary optic neuropathy. Brain 2001;124:209–218.

15 Chinnery PF, Andrews RM, Turnbull DM, Howell N: Leber hereditary optic neuropathy: Does heteroplasmy influence the inheritance and expression of the G11778A mitochondrial DNA mutation? Am J Med Genet 2001;98:235–243.

16 Johns DR: Genotype-specific phenotypes in Leber's hereditary optic neuropathy. Clin Neurosci 1994;2:146–150.

17 Nakamura M, Yamamoto M: Variable pattern of visual recovery of Leber's hereditary optic neuropathy. Br J Ophthalmol 2000;84:534–535.

18 Shoffner JM, Brown MD, Stugard C, Jun AS, Pollock S, Haas RH, Kaufman A, Koontz D, Kim Y: Leber's hereditary optic neuropathy plus dystonia is caused by a mitochondrial DNA point mutation. Ann Neurol 1995;38:163–169.

19 Kirby DM, Kahler SG, Freckmann ML, Reddihough D, Thorburn DR: Leigh disease caused by the mitochondrial DNA G14459A mutation in unrelated families. Ann Neurol 2000;48:102–104.

20 De Vries DD, Went LN, Bruyn GW, Scholte HR, Hofstra RMW, Bolhuis PA, van Oost BA: Genetic and biochemical impairment of mitochondrial complex I activity in a family with Leber hereditary optic neuropathy and hereditary spastic dystonia. Am J Hum Genet 1996;58:703–711.
21 Howell N: Primary LHON mutations: Trying to separate 'fruyt' from 'chaf'. Clin Neurosci 1994; 2:130–137.
22 Pulkes T, Eunson L, Patterson V, Siddiqui A, Wood NW, Nelson IP, Morgan-Hughes JA, Hanna MG: The mitochondrial DNA G13513A transition in ND5 is associated with a LHON/MELAS overlap syndrome and may be a frequent cause of MELAS. Ann Neurol 1999;46:916–919.
23 Corona P, Antozzi C, Carrara F, D'Incerti L, Lamantea E, Tiranti V, Zeviani M: A novel mtDNA mutation in the ND5 subunit of complex I in two MELAS patients. Ann Neurol 2001;49:106–110.
24 Howell N, Miller NR, Mackey DA, Arnold A, Herrnstadt C, Williams IM, Kubacka I: Lightning strikes twice: Leber hereditary optic neuropathy families with two pathogenic mtDNA mutations. J Neuroophthalmol 2002;22:262–269.
25 Brown MD, Sun F, Wallace DC: Clustering of Caucasian Leber hereditary optic neuropathy patients containing the 11778 or 14484 mutations on an mtDNA lineage. Am J Hum Genet 1997;60:381–387.
26 Torroni A, Petrozzi M, D'Urbano L, Sellitto D, Zeviani M, Carrara F, Carducci C, Leuzzi V, Carelli V, Barboni P, De Negri A, Scozzari R: Haplotype and phylogenetic analyses suggest that one European-specific mtDNA background plays a role in the expression of Leber hereditary optic neuropathy by increasing the penetrance of the primary mutations 11778 and 14484. Am J Hum Genet 1997;60:1107–1121.
27 Brown MD: The enigmatic relationship between mitochondrial dysfunction and Leber's hereditary optic neuropathy. J Neurol Sci 1999;165:1–5.
28 Brown MD, Trounce IA, Jun AS, Allen JC, Wallace DC: Functional analysis of lymphoblast and cybrid mitochondria containing the 3460, 11778, or 14484 Leber's hereditary optic neuropathy mitochondrial DNA mutation. J Biol Chem 2000;275:39831–39836.
29 Carelli V, Ghelli A, Bucchi L, Montagna P, De Negri A, Leuzzi V, Carducci C, Lenaz G, Lugaresi E, Degli Esposti M: Biochemical features of mtDNA 14484 (ND6/M64V) point mutation associated with Leber's hereditary optic neuropathy. Ann Neurol 1999;45:320–328.
30 Cock HR, Cooper JM, Schapira AHV: Functional consequences of the 3460-bp mitochondrial DNA mutation associated with Leber's hereditary optic neuropathy. J Neurol Sci 1999;165:10–17.
31 Lodi R, Taylor DJ, Tabrizi SJ, Kumar S, Sweeney M, Wood NW, Styles P, Radda GK, Schapira AHV: In vivo skeletal muscle mitochondrial function in Leber's hereditary optic neuropathy assessed by ^{31}P magnetic resonance spectroscopy. Ann Neurol 1997;42:573–579.
32 Lodi R, Montagna P, Cortelli P, Iotti S, Cevoli S, Carelli V, Barbiroli B: 'Secondary' 4216/ND1 and 13708/ND5 Leber's hereditary optic neuropathy mitochondrial DNA mutations do not further impair in vivo mitochondrial oxidative metabolism when associated with the 11778/ND4 mitochondrial DNA mutation. Brain 2000;123:1896–1902.
33 Lodi R, Carelli V, Cortelli P, Iotti S, Valentino ML, Barboni P, Palloti F, Montagna P, Barbiroli B: Phosphorus MR spectroscopy shows a tissue specific in vivo distribution of biochemical expression of the G3460A mutation Leber's hereditary optic neuropathy. J Neurol Neurosurg Psychiatry 2002;72:805–807.
34 Klivenyi P, Karg E, Rozsa C, Horvath R, Komoly S, Nemeth I, Turi S, Vecsei L: α-Tocopherol/lipid ratio in blood is decreased in patients with Leber's hereditary optic neuropathy and asymptomatic carriers of the 11778 mtDNA mutations. J Neurol Neurosurg Psychiatry 2001;70: 359–362.
35 Wong A, Cortopassi G: mtDNA mutations confer cellular sensitivity to oxidant stress that is partially rescued by calcium depletion and cyclosporin A. Biochem Biophys Res Commun 1997;234:511–515.
36 Danielson SR, Wong A, Carelli V, Martinuzzi A, Schapira AHV, Cortopassi GA: Cells bearing mutations causing Leber's hereditary optic neuropathy are sensitized to Fas-induced apoptosis. J Biol Chem 2002;277:5810–5815.
37 Wong A, Cavelier L, Collins-Schramm HE, Seldin MF, McGrogan M, Savontaus ML, Cortopassi GA: Differentiation-specific effects of LHON mutations introduced into neuronal cells. Hum Mol Genet 2002;11:431–438.

38 Kortuem K, Geiger LK, Levin LA: Differential susceptibility of retinal ganglion cells to reactive oxygen species. Invest Ophthalmol Vis Sci 2000;41:3176–3182.
39 Wang J, Green PS, Simpkins JW: Estradiol protects against ATP depletion, mitochondrial membrane potential decline and the generation of reactive oxygen species induced by 3-nitropropionic acid in SK-N-SH human neuroblastoma cells. J Neurochem 2001;77:804–811.
40 Johnston PB, Gaster RN, Smith VC, Tripathi RC: A clinico-pathological study of autosomal dominant optic atrophy. Am J Ophthalmol 1979;88:868–875.
41 Votruba M, Fitzke FW, Holder GE, Carter A, Bhattacharya SS, Moore AT: Clinical features in affected individuals from 21 pedigrees with dominant optic atrophy. Arch Ophthalmol 1998;116: 351–358.
42 Jacobson DM, Stone EM: Difficulty differentiating Leber's from dominant optic neuropathy in a patient with remote visual loss. J Clin Neuroophthalmol 1991;11:152–157.
43 Toomes C, Marchbank NJ, Mackey DA, Craig JE, Newbury-Ecob RA, Bennett CP, Vize CJ, Desai SP, Black GCM, Patel N, Teimory M, Markham AF, Inglehearn CF, Churchill AJ: Spectrum, frequency and penetrance of *OPA1* mutations in dominant optic atrophy. Hum Mol Genet 2001;10:1369–1378.
44 Alexander C, Votruba M, Pesch UEA, Thiselton DL, Mayer S, Moore A, Rodriguez M, Kellner U, Leo-Kottler B, Auburger G, Bhattacharya SS, Wissinger B: *OPA1*, encoding a dynamin-related GTPase, is mutated in autosomal dominant optic atrophy linked to chromosome 3q28. Nat Genet 2000;26:211–215.
45 Delettre C, Lenaers G, Griffoin JM, Gigarel N, Lorenzo C, Belenguer P, Pelloquin L, Grosgeorge J, Turc-Carel C, Perret E, Astarie-Dequeker C, Lasquellec L, Arnaud B, Ducommun B, Kaplan J, Hamel CP: Nuclear gene *OPA1*, encoding a mitochondrial dynamin-related protein, is mutated in dominant optic atrophy. Nat Genet 2000;26:207–210.
46 Pesch UEA, Leo-Kottler B, Mayer S, Jurklies B, Kellner U, Apfelstedt-Sylla E, Zrenner E, Alexander C, Wissinger B: *OPA1* mutations in patients with autosomal dominant optic atrophy and evidence for semi-dominant inheritance. Hum Mol Genet 2001;10:1359–1368.
47 Thiselton DL, Alexander C, Taanman JW, Brooks S, Rosenberg T, Eiberg H, Andreasson S, Van Regemorter N, Munier FL, Moore AT, Bhattacharya SS, Votruba M: A comprehensive survey of mutations in the *OPA1* gene in patients with autosomal dominant optic atrophy. Invest Ophthalmol Vis Sci 2002;43:1715–1724.
48 Tatton WG, Chalmers-Redman RME, Sud A, Podos SM, Mittag TW: Maintaining mitochondrial membrane impermeability: An opportunity for new therapy in glaucoma? Surv Ophthalmol 2001;45(suppl 3):277–283.
49 Aung T, Ocaka L, Ebenezer ND, Morris AG, Krawczak M, Thiselton DL, Alexander C, Votruba M, Brice G, Child AH, Francis PJ, Hitchings RA, Lehmann OJ, Bhattacharya SS: A major marker for normal tension glaucoma: association with polymorphisms in the *OPA1* gene. Hum Hered 2002;110:52–56.
50 Opial D, Boehnke M, Tadesse S, Lietz-Partzsch A, Flammer J, Munier F, Mermoud A, Hirano M, Fluckiger F, Mojon DS: Leber's hereditary optic neuropathy mitochondrial DNA mutations in normal-tension glaucoma. Graefes Arch Clin Exp Ophthalmol 2001:239:437–440.
51 Fingert JH, Heon E, Liebmann JM, Yamamoto T, Craig JE, Rait J, Kawase K, Hoh ST, Buys YM, Dickinson J, Hockey RR, Williams-Lyn D, Trope G, Kitazawa Y, Ritch R, Mackey DA, Alward WLM, Sheffield, Stone EM: Analysis of myocilin mutations in 1703 glaucoma patients from five different populations. Hum Mol Genet 1999;8:899–905.
52 Shimizu S, Lichter PR, Johnson AT, Zhou Z, Higashi M, Gottfredsdottir M, Othman M, Moroi SE, Rozsa FW, Schertzer RM, Clarke MS, Schwartz AL, Downs CA, Vollrath D, Richards JE: Age-dependent prevalence of mutations at the *GLC1* locus in primary open-angle glaucoma. Am J Ophthalmol 2000;130:165–177.
53 Faucher M, Anctil JL, Rodrigue MA, Duchesne A, Bergeron D, Blondeau P, Côte G, Dubois S, Bergeron J, Arseneault R, The Québec Glaucoma Network, Morissette J, Raymond V: Founder *TIGR/myocilin* mutations for glaucoma in the Québec population. Hum Mol Genet 2002;11:2077–2090.
54 Wentz-Hunter K, Ueda J, Shimizu N, Yue BYJT: Myocilin is associated with mitochondria in human trabecular meshwork cells. J Cell Physiol 2002;190:46–53.

55 Rezaie T, Child A, Hitchings R, Brice G, Miller L, Coca-Prados M, Heon E, Krupin T, Ritch R, Kreutzer D, Crick RP, Sarfarazi M: Adult-onset primary open-angle glaucoma caused by mutations in optineurin. Science 2002;295:1077–1079.
56 Cechetto JD, Gupta RS: Immunoelectron microscopy provides evidence that tumor necrosis factor receptor-associated protein 1 is a mitochondrial protein which also localizes at specific extramitochondrial sites. Exp Cell Res 2000;260:30–39.
57 Ledgerwood EC, Prins JB, Bright NA, Johnson DR, Wolfreys K, Pober JS, O'Rahilly S, Bradley JR: Tumor necrosis factor is delivered to mitochondria where a tumor necrosis factor-binding protein is localized. Lab Invest 1998;78:1583–1589.
58 Heng JE, Vorwerk CK, Lessell E, Zurakowski D, Levin LA, Dreyer EB: Ethambutol is toxic to retinal ganglion cells via an excitotoxic pathway. Invest Ophthalmol Vis Sci 1999;40:190–196.
59 Yoon YH, Jung KH, Sadun AA, Shin HC, Koh JY: Ethambutol-induced vacuolar changes and neuronal loss in rat retinal cell culture: Mediation by endogenous zinc. Toxicol Appl Pharmacol 2000;162:107–114.
60 Bristow EA, Griffiths PG, Andrews RM, Johnson MA, Turnbull DM: The distribution of mitochondrial activity in relation to optic nerve structure. Arch Ophthalmol 2002;120:791–796.
61 Szewczyk A, Wojtczak L: Mitochondria as a pharmacological target. Pharmacol Rev 2002; 54:101–127.

Dr. Neil Howell
MitoKor, 11494 Sorrento Valley Road,
San Diego, CA 92121 (USA)
Tel. +1 858 7937800, Fax +1 858 7937805, E-Mail howelln@mitokor.com

Wissinger B, Kohl S, Langenbeck U (eds): Genetics in Ophthalmology.
Dev Ophthalmol. Basel, Karger, 2003, vol 37, pp 109–125

Retinitis pigmentosa: Genes, Proteins and Prospects

M.M. Hims[a], *S.P. Diager*[b], *C.F. Inglehearn*[a]

[a] Molecular Medicine Unit, St James's University Hospital, Leeds, UK and
[b] Human Genetics Center, School of Public Health,
The University of Texas Health Science Center, Houston, Tex., USA

Abstract

The name retinitis pigmentosa (RP) describes a heterogeneous group of inherited progressive retinal dystrophies, primarily affecting the peripheral retina. Patients experience night blindness and visual field loss, often leading to complete blindness. RP can be inherited in autosomal dominant, autosomal recessive, X-linked, mitochondrial and genetically more complex modes. To date, 39 loci have been implicated in non-syndromic RP, for which 30 of the genes are known. Many of these can be grouped by function, giving insights into the disease process. These include components of the phototransduction cascade, proteins involved in retinol metabolism and cell-cell interaction, photoreceptor structural proteins and transcription factors, intracellular transport proteins and splicing factors. Current knowledge of each grouping is reviewed briefly herein and consistent patterns of inheritance, which may have functional significance, are noted. The complexity of these diseases has in the past made it difficult to counsel patients or to envisage widely applicable therapies. As a more complete picture is emerging however, possibilities exist for streamlining screening services and a number of avenues for possible therapy are being investigated.

In 1857, shortly after the invention of the ophthalmoscope, the German physician Donders observed 'bone spicule' pigmentation of the retina in some forms of blindness [1]. To describe what he saw, Donders coined the term retinitis pigmentosa (RP), technically a misnomer since the primary defect is not inflammatory, but the name has stuck and is now widely used. In its modern usage, RP describes a heterogeneous group of progressive retinal dystrophies, primarily affecting the peripheral retina. Patients experience night blindness and progressive loss of visual fields, leading to complete blindness in around

30% of cases and severe visual disability in the remainder. It is the commonest inherited retinal dystrophy, affecting approximately 1 in 3,500 people or around 2 million sufferers world-wide [2, 3]. Ophthalmic examination reveals constriction of retinal arterioles, optic disc pallor and characteristic bone spicule-like pigmentary deposits, while electrodiagnostic testing shows a reduced or abolished rod electroretinogram [4]. The pigmentation in RP is thought to result from either the migration of pigment-containing retinal pigment epithelial (RPE) cells into the degenerating retina or from macrophages travelling from the choriocapillaris to the damaged retina, picking up pigmentation as they pass through the degenerating RPE.

RP can be inherited in autosomal dominant, autosomal recessive, X-linked and rare mitochondrial and digenic forms. However approximately 50% of RP patients have no family history of the disease [5]. These are generally assumed to be recessive in nature, though they could be cases of X-linked or dominant RP with partial penetrance, or new dominant mutations. It is also possible that a proportion of these cases represent more genetically complex disease resulting from unfavourable alleles at multiple loci.

The last decade has seen rapid advancement in our understanding of the genetic basis of RP. Through such studies it has become apparent that RP displays unprecedented genetic heterogeneity. Here we review current knowledge of the causes of RP occurring in the absence of other defects. Syndromic RP and other forms of retinal dystrophy are reviewed elsewhere in this volume. To date, 39 loci have been implicated in the typical forms of non-syndromic RP, and the disease-causing gene has been identified in 30 of these cases, with genes at 9 loci (1 dominant, 5 recessive and 3 X-linked) remaining to be identified. However, RP is, in fact, just one of a number of related retinal degenerative diseases. A further 33 genes or loci have been implicated in various syndromic forms of RP and in total, 132 genes or loci have been shown to underlie all the different forms of human retinal dystrophies [6].

With a large number of causative genes now identified it has become possible to group most of the known RP proteins into six tentative functional classes. Table 1 summarizes these groups and subsequent sections describe the current understanding of each. Grouping the genes in this fashion gives some insight into the types of cellular defects that lead to retinal degeneration.

The Phototransduction Cascade

Phototransduction is the well-characterized biochemical process that converts a photon of light into an electrochemical signal within the photoreceptors (reviewed by Molday [7]). Rods and cones use the same basic mechanism,

Table 1. A summary of the genes implicated in RP, organized into functional groups

Gene	Protein	Associated dystrophies
Phototransduction cascade		
RHO	Rhodopsin	adRP, arRP and adCSNB
PDE6A	Phosphodiesterase α subunit	arRP
PDE6B	Phosphodiesterase β subunit	arRP
SAG	Arrestin	arRP, arCSNB
CNGA1	Rod cGMP-gated channel α subunit	arRP
CNGB1	Rod cGMP-gated channel β subunit	arRP
Visual cycle		
RPE65	Retinal pigment epithelium-specific 65-kd protein	arRP, LCA
RLBP	Cellular retinaldehyde-binding protein	arRP and recessive retinitis punctata albescens
ABCA4	ATP-binding cassette transporter	arRP, CRD and macular dystrophy, ARMD?
LRAT	Lecithin retinol acyltransferase	arRP
RGR	RPE-retinal G protein-coupled receptor	arRP and dominant choroidal sclerosis
Structural proteins		
RDS	Peripherin/RDS	adRP, digenic RP, ad macular/pattern dystrophy
ROM1	Retinal outer segment protein 1	digenic RP
RHO	Rhodopsin	adRP, arRP and adCSNB
Transcription factors		
CRX	Cone-rod otx-like homeobox transcription factor	adRP, LCA, adCRD
NRL	Neural retinal leucine zipper	adRP
NR2E3	Nuclear receptor subfamily 2 group E3	arRP, enhanced S-cone syndrome
Splicing factors		
PRPF8	Pre-mRNA processing factor 8	adRP
PRPF3	Pre-mRNA processing factor 3	adRP
PRPF31	Pre-mRNA processing factor 31	adRP

Table 1 (continued)

Gene	Protein	Associated dystrophies
Intracellular transport		
FSCN2	Retinal fascin	adRP
RP1	RP1 protein	adRP
RPGR	Retinitis pigmentosa GTPase regulator	X-linked RP
RP2	RP2 protein	X-linked RP
TULP1	Tubby-like protein 1	arRP
Cell-cell adhesion/signalling		
MERTK	C-mer proto-oncogene receptor tyrosine kinase	arRP
CRB1	Crumbs homolog 1	arRP, LCA
USH2A	Usherin	arRP, Ushers syndrome
Miscellaneous		
IMPDH1	Inosine monophosphate dehydrogenase 1	adRP
RP9	RP9 protein	adRP

namely a G-protein-coupled receptor signalling pathway, although distinct rod and cone homologs exist for many of the components of the pathway. Since RP is primarily a disease of the rods, it is not surprising that mutations in many of the components of rod phototransduction were among the first to be implicated in RP.

The phototransduction cascade in rod photoreceptor cells (reviewed in figure 1) is initiated by rhodopsin, a seven-pass transmembrane protein covalently linked to an 11-*cis* retinal chromophore. Photoexcitation converts 11-*cis* retinal to its all-*trans* isomer, creating Meta II rhodopsin, which catalyses the activation of the G-protein transducin. This in turn leads to the activation of phosphodiesterase (PDE), which hydrolyses cGMP to 5′-GMP. One rhodopsin molecule can activate multiple transducin molecules, which in turn activate multiple PDE molecules, amplifying the signal. The decrease in intracellular cGMP causes the cGMP-gated cation channels in the outer segment membrane to close. Without the balanced influx of Ca^{2+} the cell becomes hyperpolarized. In this state the release of glutamate transmitter from the synaptic region is inhibited and a signal is sent to the visual cortex. The resting phase is restored when Meta II rhodopsin is inactivated by rhodopsin kinase and binds to arrestin. Transducin and PDE are inactivated and disassociate due to the hydrolysis of the bound GTP by the intrinsic GTPase activity of the transducin α subunit. The low

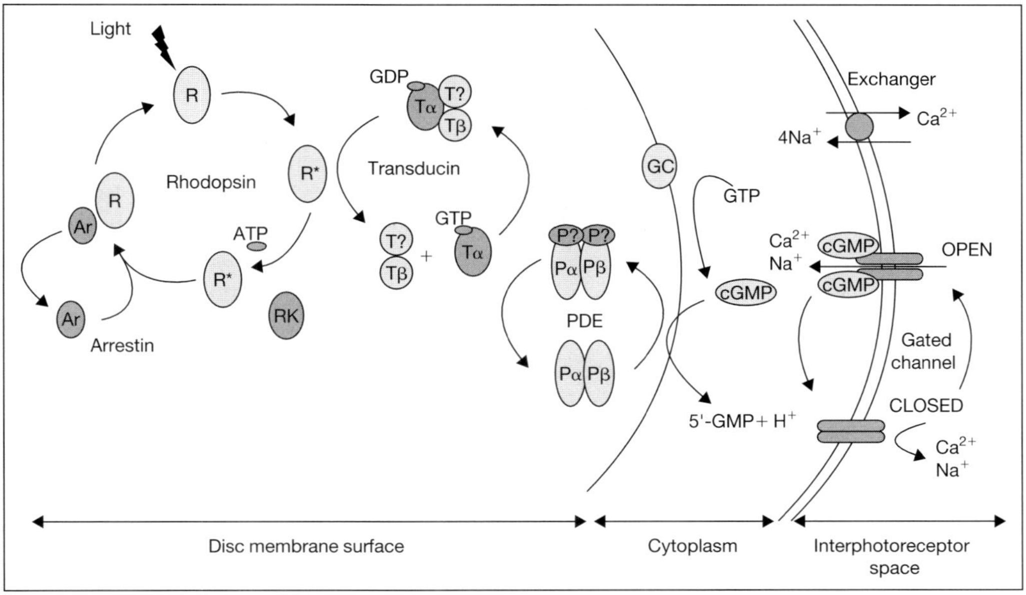

Fig. 1. Major components and key events of the phototransduction cascade. R = Rhodopsin, R* = activated Meta II rhodopsin, RK = rhodopsin kinase, Ar = arrestin, T = transducin, P = phosphodiesterase (PDE), GC = guanylate cyclase [adapted from 8].

level of intracellular Ca^{2+} caused by the closure of the cGMP-gated channels activates guanylate cyclase, which synthesizes cGMP. As intracellular levels of cGMP increase, the cGMP-gated Na^{+} and Ca^{2+} channels reopen and a depolarized dark state is re-established. As Ca^{2+} levels rise again, guanylate cyclase activity is inhibited and cGMP synthesis returns to basal levels.

Mutations in rhodopsin (*RHO*) are the most common single cause of RP and over 100 disease-causing mutations have been reported [9]. These primarily cause dominant RP, but mutations causing recessive RP and dominant congenital stationary night blindness (CSNB) have also been reported. Mutations in genes encoding the rod α and β subunits of PDE (*PDE6A* and *PDE6B*), the subunit of the rod cyclic nucleotide-gated channel (*CNGA1*), and arrestin (*SAG*) all cause recessive RP [6]. A subset of mutations in *PDE6B* can, in addition, cause dominant CSNB, while certain mutations in *SAG* cause recessive CSNB. Components of the phototransduction cascade have also been implicated in other retinal dystrophies. These include retina-specific guanylate cyclase (*GUCY2D)* in cone-rod dystrophy or Leber's amaurosis, guanyate cyclase-activating protein 1A (*GUCA1A)* in cone dystrophy, rhodopsin kinase (*RHOK*)

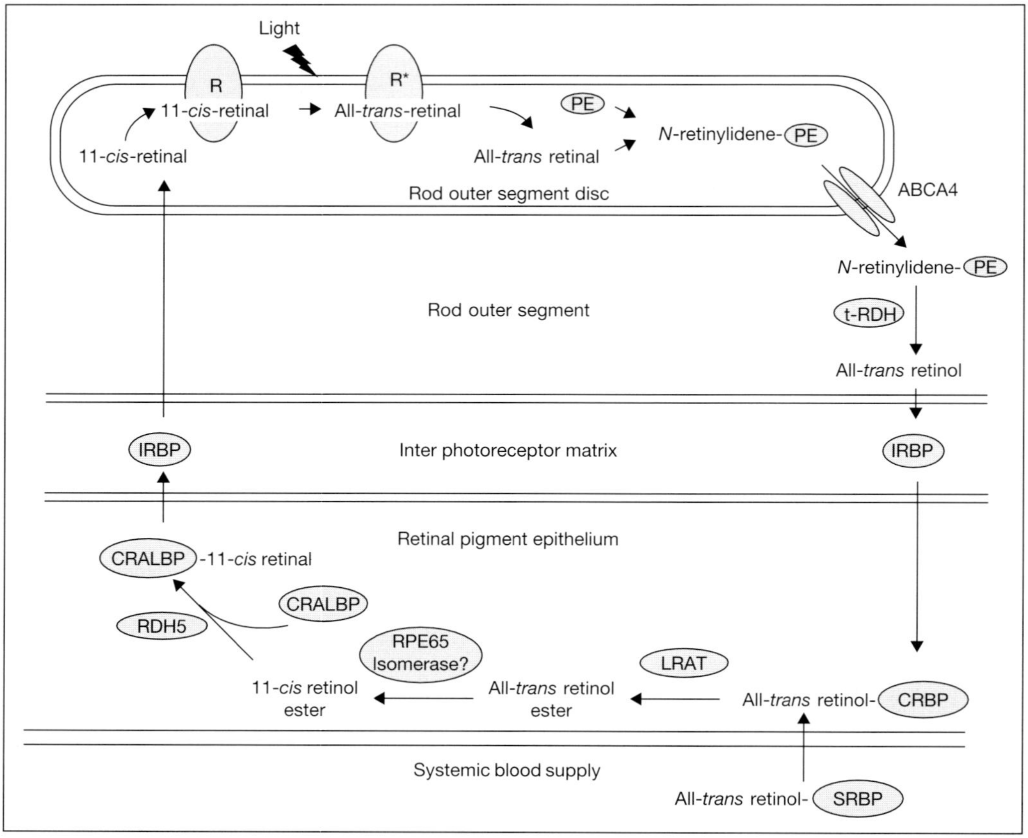

Fig. 2. The major known components of the visual cycle. R = Rhodopsin, ABCA4 = ATP-binding cassette transporter, RPE65 = retinal pigment epithelium 65 kd protein, RDH = retinol dehydrogenase, RDH5 = 11-*cis* retinol dehydrogenase 5, LRAT = lecithin retinol acyltransferase, SRBP = serum retinol-binding protein, IRBP = interphotoreceptor retinoid-binding protein, CRALBP = cellular retinaldehyde-binding protein, PE = phosphatidylethanolamine [adapted from 10].

and rod transducin α subunit (*GNAT1*) in recessive CSNB, and the cone cyclic nucleotide-gated channel (*CNGA3*) in achromatopsia [6].

Retinol (Vitamin A) Metabolism

The visual cycle is the name given to a series of biochemical steps that recycle the chromophore component of rhodopsin, as shown in figure 2.

Elsewhere in this volume Thompson and Gal comprehensively review this process and its implications for RP pathogenesis. The following is an overview of the visual cycle and of those components of it involved in RP causation.

Inside the RPE, all-*trans* retinol is bound to cytosolic retinoid-binding protein and is esterified by lecithin retinol transferase (LRAT) to produce an all-*trans* retinol ester. The hydrolysis of the ester bond by an isomerohydrolase provides the energy for the isomerization of the all-*trans* retinol to 11-*cis* retinol, in a process believed to involve the RPE protein RPE65. 11-*cis* retinol is then bound by cellular retinaldehyde-binding protein (CRALBP) before being converted to 11-*cis* retinal by the action of 11-*cis* dehydrogenase 5 (RDH5). The 11-*cis* retinal molecule is transported to the photoreceptor where it is bound to rhodopsin by a protonated Schiff base linkage. Photoexcitation of rhodopsin converts the 11-*cis* retinal to all-*trans* retinal and initiates the phototransduction cascade. The all-*trans* retinal is released from rhodopsin and binds to phosphatidylethanolamine (PE) to form *N*-retinylidene-phosphatidylethanolamine (*N*-retinylidene-PE). This substrate is then thought to be transported from the interior of the discs into the photoreceptor cytoplasm by the ATP-binding cassette transporter (ABCA4). Within the cytoplasm, *N*-retinylidene-PE is converted to all-*trans* retinol by all-*trans* retinol dehydrogenase and released for transportation to the RPE, completing the cycle.

Mutations causing recessive RP have been found in four genes encoding components of the visual cycle, namely *RPE65*, *CRALBP*, *LRAT* and *ABCA4*. Both *LRAT* and *RPE65* mutations lead to a particularly severe early onset form of RP, while certain mutations in *ABCA4* can also cause a range of central retinopathies. Mutations causing recessive RP have also been identified in the gene encoding the RPE-retinal G-protein-coupled receptor (*RGR*), a homolog of rhodopsin that facilitates the photoisomerization of all-*trans* retinal to 11-*cis* retinal (the reverse of the reaction that occurs upon photoisomerization of rhodopsin). The exact role of this protein in retinal biology is uncertain but it has been proposed that it provides an alternative source of chromophore under conditions of high light flux.

Structural Proteins

The photoreceptor-specific proteins peripherin/RDS and ROM1 are thought to be structural proteins vital for maintaining the shape of the flattened discs of the rod outer segment. These proteins both have a 4-transmembrane domain structure and have been found to form heterotetramers with each other at the rims of discs [7]. Mutations in *RDS* have been implicated in dominant RP and a digenic form of RP that requires heterozygous mutations in both the *RDS* and *ROM1* [11]. Mutations in *ROM1* alone have only been tentatively implicated in dominant RP [12]. Allelic mutations in peripherin/RDS can also cause a variety of central retinal dystrophies [13].

Rhodopsin accounts for an estimated 70% of the photoreceptor outer segment protein content [14]. It has therefore been suggested that, in addition to its role in phototransduction, rhodopsin may also have an important structural role in the morphology of the rod outer segments [15].

Transcription Factors

The transcription factors encoded by *CRX*, *NRL* and *NR2E3* are all thought to be involved in the control of many photoreceptor-specific genes [16–18]. Mutations in *NRL* have so far been found only in dominant RP [19], while mutations in *CRX* have been detected in a range of phenotypes including dominant RP, cone-rod dystrophy and LCA [20]. Mutations in *NR2E3* have been identified in recessive RP patients [21] as well as in the rare recessive retinopathy enhanced S-cone syndrome [22].

Cell-Cell Interactions

The C-mer proto-oncogene tyrosine kinase (*MERTK*), mutated in cases of recessive RP, is thought to have a role in cell-cell signalling between the RPE and the photoreceptor cells [23]. The gene *CRB1*, encoding the transmembrane protein crumbs, has recently been shown in *Drosophila* experiments to be essential for correct photoreceptor morphogenesis, playing a role in positional signalling and in control of the distribution of adherin proteins between photoreceptor cells [24]. CRB1 has been implicated in both RP and LCA phenotypes [25, 26]. In addition, the usherin gene *USH2A*, thought to be a cell-cell adhesion molecule, has been found to be mutated in cases of recessive RP [27]. Usherin is one of three proteins with predicted cell adhesion properties that have been implicated in Usher syndrome, the other two proteins being the PCDH15 and CDH23, both members of the cadherin protein family [28, 29].

Splicing Factors

One of the most intriguing discoveries over the last 12 months in the field of RP genetics has been the identification of mutations in three pre-mRNA splicing factors in autosomal dominant RP patients. The implicated genes *PRPF8*, *PRPF31* and *HPRP3* all encode small nuclear ribonucleoprotein (snRNP) proteins that are essential components of the U4/U6:U5 tri-snRNP, itself a major component of the spliceosome [30–32]. The spliceosome is a large multimeric RNA-protein complex, which functions to remove intronic sequences from pre-mRNA transcripts and splice together the exons to produce mature mRNA [33]. The basic components of the spliceosomal splicing mechanism are shown in figure 3. The three RP-implicated splicing factor genes are all expressed ubiquitously, unlike other RP genes, the majority of which have retina-specific expression patterns. Exactly why defects in a basic housekeeping cellular function should cause a dominant retina-specific disease is not known. It is possible that the high metabolic rate of retinal cells and the need to continually synthesize large amounts of photopigment make photoreceptors particularly vulnerable to defects which are sub-pathological in other tissues. Alternatively, the processing of one or several retina-specific transcripts may be uniquely affected by these mutations.

Intracellular Transport/Cytoskeleton Function

With the photoreceptors showing such a highly structured morphology, the efficient transportation of proteins from the site of synthesis in the inner segment to the outer segment via the cilium would be expected to be essential to photoreceptor function. The product of the gene *TULP1*, a retina-specific member of the tubby-like gene family, is thought to play a role in the transport of rhodopsin from the inner to the outer segment [34]. The product of the *RPGR* gene, mutations in which account for 70% of X-linked of RP, co-localizes to the outer segment and cilium of the photoreceptors and is proposed to be involved in vesicular transport [35]. Furthermore, three of the genes defective in RP patients appear to have a role in the formation of the cell cytoskeleton. The integral role of the cytoskeleton in intracellular transport suggests a possible mechanism by which all three of these mutated proteins may exert a pathogenic effect. The *RP1* gene is mutated in dominant RP and shows homology to doublecortin (*DCX*), which is believed to have a function in the regulation of microtubule dynamics and stability during neuronal development [36]. The RP1 protein is present only in the retina and has been shown to localize to the cilium of the photoreceptor [37]. The *RP2* gene is mutated in some forms of X-linked RP and shares homology to the human cofactor C, which is involved in β-tubulin folding [38]. Finally, the retinal specific gene *FSCN2*, which has been tentatively associated with dominant RP, is believed to play a role in actin bundling [39].

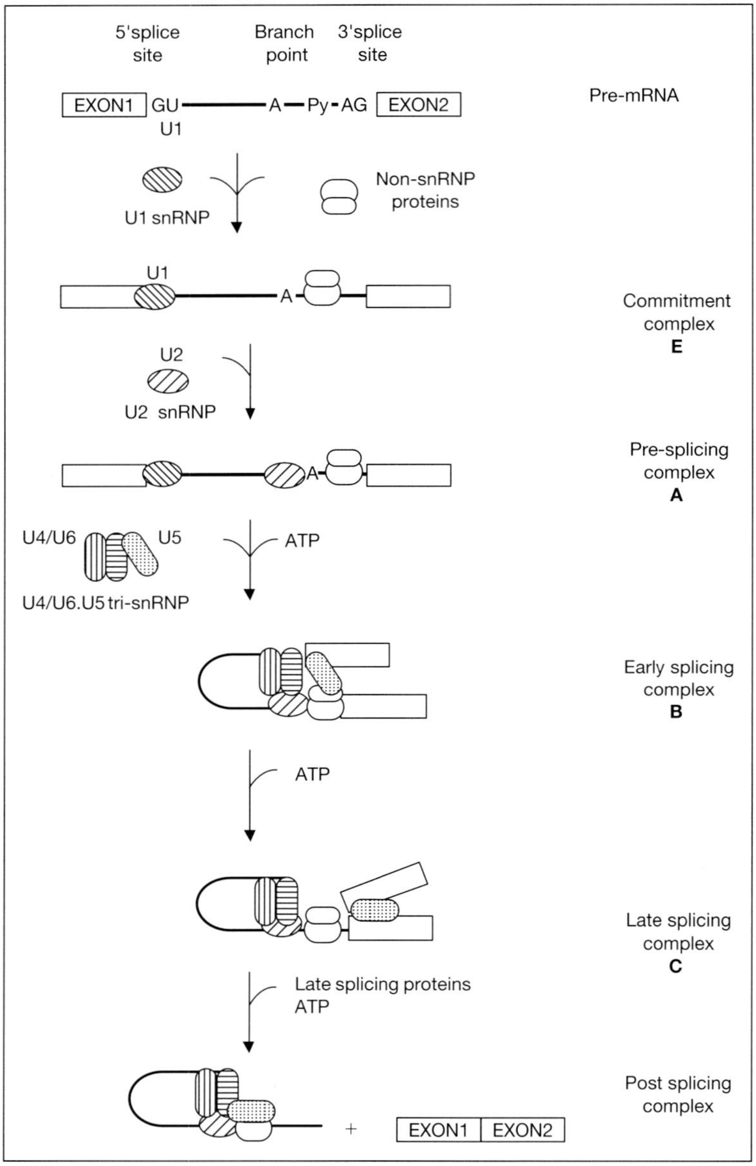

Fig. 3. Diagram showing the major known components of spliceosome assembly pathway. The snRNP proteins are represented by shaded symbols and the non-snRNP proteins by clear symbols. The proteins encoded by PRPF8, PRPF31 and HPRP3 are all part of the U4/U6.U5snSNP complex [adapted from 33].

Genes of Unknown Function

Two recently identified RP genes have less well-defined roles in retinal biology and cannot be placed into one of the above functional categories. These are inosine monophosphate dehydrogenase 1(*IMPDH1*), which encodes a protein involved in nucleotide biosynthesis [40, 41], and retinitis pigmentosa 9 (*RP9*) a ubiquitously expressed gene of unknown function [42]. It remains to be determined by what mechanism mutations in either of these genes manifest themselves as an RP phenotype.

Pathways to Retinal Degeneration

Many of the mutations described above might be expected to reduce or abolish retinal function. Instead, in RP the retina develops and functions normally, and vision is lost only later in the disease process by progressive retinal degeneration. However, in the retinal dystrophy CSNB, patients lack rod function from birth but do not progress to RP. It is interesting to note that CSNB-causing mutations are allelic with mutations leading to RP in genes encoding 3 components of the phototransduction cascade. A fourth gene, *RHOK*, has been implicated in dominant CSNB alone. Furthermore, defects in components of two cone phototransduction proteins the cone cGMP-gated channel α and β subunits, *CNGA3* and *CNGB3*, cause achromatopsia, a non-progressive form of colour blindness. It therefore appears that some defects in phototransduction cause retinal degeneration but others cause non-progressive loss of photoreceptor function.

Another pattern that may have functional significance is that of mode of inheritance. Defects within the phototransduction cascade almost always lead to recessive RP. The one notable exception, rhodopsin, could in fact have an additional structural role, and indeed both of the known photoreceptor structural proteins implicated in RP cause a dominant disease. Defects in the visual cycle components are consistently recessive in effect, while mutations in retinal transcription factors and in components of the spliceosome cause dominant RP. Only within the tentatively grouped 'intracellular transport' category is there a mix of inheritance types. Within this group, mutations in the two proteins putatively involved in cytoskeleton formation cause dominant RP, *FSCN1* and *RP1*, while the proteins with a more direct role in cellular transport, *RPGR1*, *RP2* and *TULP1*, are implicated in recessive or X-linked disease. This may reflect a further subdivision within this group. For many RP genes though, a minority of mutations exist that do not fit the expected pattern.

The finding of defects in these pathways and processes leading to RP provides a significant insight into RP pathogenesis, but many questions still remain. In a number of animal models of retinal degeneration it has been shown that photoreceptors die by an apoptotic pathway [43–44]. It therefore seems likely that, in most, if not all, forms of human retinal degeneration, the photoreceptors also die by apoptosis [45]. How then do this diverse set of defects lead to the common endpoint of photoreceptor apoptosis? Based on observations in animal models of retinal degenerations and our knowledge of retinal biology, several mechanisms have been proposed and each may apply to a proportion of RP cases. These mechanisms are reviewed in more detail elsewhere [45, 46], and are covered briefly below.

First, mutations that disrupt the phototransduction cascade may lead to a situation wherein the cGMP-gated channels are held open permanently. The constant influx of Na^+ and Ca^{2+} ions may cause metabolic overload and direct toxicity. Second, some mutations disrupt the formation of outer segment discs, either directly through a defect in a structural protein or indirectly due to failure of intracellular transport of integral proteins, possibly causing the ectopic accumulation of these proteins. A malformation of the outer segment discs and a shortening of the outer segment is observed in several animal models of retinal degeneration. It has been proposed that this may lead to toxic levels of oxygen, due to reduced oxygen consumption and the closer physical proximity of the photoreceptor cell body to the oxygen-rich choroid. Third, RPE dysfunction may cause retinal degenerations, by disrupting either the recycling of 11-*cis* retinal or some other essential RPE photoreceptor interaction. A fourth potential mechanism is one in which mutations lead to a constitutive activation of the phototransduction cascade. How this would trigger cell death is unknown, however, it may again involve oxygen toxicity, since oxygen consumption in the photoreceptor microenvironment would be reduced due to the unloading of the Na^+/K^+-ATPases in the constitutively activated photoreceptor. This mechanism is known as the 'equivalent light hypothesis' [47].

The level of genetic complexity of human eye diseases, in general, and of inherited retinal degenerations, in particular, is unique and unprecedented. Approximately one third of all human inherited diseases include defects of the eye [48]. This may be due in part to the non-lethality of eye defects and the obvious presentation of a retinal disorder. However, the retina is a complex, specialized, non-dividing tissue with high oxygen consumption and an unusually large number of mitochondria, implying a high metabolic rate [49]. The complex inter-relationships between rods, cones, the RPE and other retinal cells add to their vulnerability as failures in any of these cells can trigger photoreceptor death in adjacent cells [50]. It has therefore been hypothesized that the eye is uniquely sensitive to some defects that are sub-pathological elsewhere in the body. In particular, this hypothesis may help explain why defects in ubiquitous processes such as splicing lead to RP.

Future Prospects

One obvious benefit for patients from research on RP genetics is accurate diagnosis, prognosis and counselling. As yet this is not widely available, since the level of genetic heterogeneity makes the provision of diagnostic services for RP a daunting task. However, it seems likely that, for X-linked and dominant loci, the known genes account for the majority of cases, at least among Americans of European origin and Europeans in whom most mutation screening has been conducted. Mutations in two dominant RP genes, rhodopsin and peripherin/RDS, account for around 35% of all dominant cases [51]. Of the remaining ten known dominant RP genes, four have mutation 'hotspots' or common mutations that account for most if not all of the mutations found therein. For X-linked RP, approximately half of all cases result from mutations in exon ORF15 of the *RPGR* gene [52], including dominant mutations that affect carrier females [53]. Recessive RP seems likely to be more complex, but even within this category the mutations 309delA in arrestin, Cys759Phe in usherin and various mutations in the catalytic domains of the α and β subunits of PDE account for a significant proportion of cases in certain populations [27, 54, 55]. Thus by selectively screening for the more common mutations and focusing on exons containing mutation hotspots, it should be possible to develop an economically viable screening service of use to many RP patients.

Research into proteomics has lagged behind the genetics revolution of the past 10 years. In the near future though, two-hybrid and other protein association assays should help piece together the pathways of normal eye development. The creation of animal models for human inherited blindness will facilitate pathological and transcriptional profiling of the various disease processes. Animal models can also be used to evaluate environmental factors and the influence of genetic modifiers, as well as allowing pre-clinical trials for possible therapies. A number of therapeutic avenues are being explored at this time. These include gene therapy, both for gene replacement [56] and for introducing therapeutic agents such as ribozymes [57], dietary therapies such as vitamin A [58], transplantation of RPE cells [59] and the so-called 'bionic eye' [60]. In addition, anti-inflammatory drugs are used to treat secondary inflammation in some patients, and dark glasses are thought to be of benefit in some cases. However, it is difficult to quantify the benefit to patients of any one therapy, since each patient tested is likely to have a different disease. One direct benefit of the new genetic knowledge is that researchers will have the opportunity to test therapies on subgroups of patients with defects in the same genes or pathways. In this way it should be possible to target therapies to those who will derive maximum benefit.

References

1 Donders, F: Beiträge zur pathologischen Anatomie des Auges. 2. Pigmentbildung in der Netzhaut. Arch Ophthalmol 1857;3:139–165.
2 Bunker CH, Berson EL, Bromley WC, Hayes RP, Roderick TH: Prevalence of retinitis pigmentosa in Maine. Am J Ophthalmol 1984;97:357–365.
3 Bundey S, Crews SJ: A study of retinitis pigmentosa in the City of Birmingham. II. Clinical and genetic heterogeneity. J Med Genet 1984;21:421–428.
4 Bird AC: Retinal photoreceptor dystrophies. Am J Ophthalmol 1995;119:543–562.
5 Jay M: Figures and fantasies: The frequencies of the different genetic forms of retinitis pigmentosa. Birth Defects Orig Artic Ser 1982;18:167–173.
6 RetNet: Retinal Information Network, vol 2001, 2001.
7 Molday RS: Photoreceptor membrane proteins, phototransduction and retinal degenerative diseases. Invest Ophthalmol Vis Sci 1998;39:2493–2513.
8 Farber DB, Danciger M: Identification of genes causing photoreceptor degenerations leading to blindness. Curr Opin Neurobiol 1997;7:666–673.
9 Gal A, Apfelstedt-Sylla E, Janecke AR, Zrenner E: Rhodopsin mutations in inherited retinal dystrophies and dysfunctions. Prog Retinal Eye Res 1996;16:51–79.
10 Perrault I, Rozet JM, Gerber S, Ghazi I, Leowski C, Ducroq D, Souied E, Dufier JL, Munnich A, Kaplan J: Leber congenital amaurosis. Mol Genet Metab 1999;68:200–208.
11 Kajiwara K, Berson EL, Dryja TP: Digenic retinitis pigmentosa due to mutations at the unlinked peripherin/RDS and ROM1 loci. Science 1994;264:1604–1608.
12 Bascom RA, Liu L, Heckenlively JR, Stone EM, McInnes RR: Mutation analysis of the ROM1 gene in retinitis pigmentosa. Hum Mol Genet 1995;4:1895–1902.
13 Keen TJ, Inglehearn CF: Mutations and polymorphisms in the human peripherin-RDS gene and their involvement in inherited retinal degeneration. Hum Mutat 1996;8:297–303.
14 Hamm HE, Bownds MD: Protein complement of rod outer segments of frog retina. Biochemistry 1986;25:4512–4523.
15 Van Soest S, Westerveld A, de Jong PT, Bleeker-Wagemakers EM, Bergen AA: Retinitis pigmentosa: Defined from a molecular point of view. Surv Ophthalmol 1999;43:321–334.
16 Kumar R, Chen S, Scheurer D, Wang QL, Duh E, Sung CH, Rehemtulla A, Swaroop A, Adler R, Zack DJ: The bZIP transcription factor Nrl stimulates rhodopsin promoter activity in primary retinal cell cultures. J Biol Chem 1996;271:29612–29618.
17 Furukawa T, Morrow EM, Cepko CL: Crx, a novel otx-like homeobox gene, shows photoreceptor-specific expression and regulates photoreceptor differentiation. Cell 1997;91:531–541.
18 Kobayashi M, Takezawa S, Hara K, Yu RT, Umesono Y, Agata K, Taniwaki M, Yasuda K, Umesono K: Identification of a photoreceptor cell-specific nuclear receptor. Proc Natl Acad Sci USA 1999;96:4814–4819.
19 Bessant DA, Payne AM, Mitton KP, Wang QL, Swain PK, Plant C, Bird AC, Zack DJ, Swaroop A, Bhattacharya SS: A mutation in NRL is associated with autosomal dominant retinitis pigmentosa. Nat Genet 1999;21:355–356.
20 Sohocki MM, Sullivan LS, Mintz-Hittner HA, Birch D, Heckenlively JR, Freund CL, McInnes RR, Daiger SP: A range of clinical phenotypes associated with mutations in CRX, a photoreceptor transcription-factor gene. Am J Hum Genet 1998;63:1307–1315.
21 Gerber S, Rozet JM, Takezawa SI, dos Santos LC, Lopes L, Gribouval O, Penet C, Perrault I, Ducroq D, Souied E, Jeanpierre M, Romana S, Frezal J, Ferraz F, Yu-Umesono R, Munnich A, Kaplan J: The photoreceptor cell-specific nuclear receptor gene (PNR) accounts for retinitis pigmentosa in the Crypto-Jews from Portugal (Marranos), survivors from the Spanish Inquisition. Hum Genet 2000;107:276–284.
22 Haider NB, Jacobson SG, Cideciyan AV, Swiderski R, Streb LM, Searby C, Beck G, Hockey R, Hanna DB, Gorman S, Duhl D, Carmi R, Bennett J, Weleber RG, Fishman GA, Wright AF, Stone EM, Sheffield VC: Mutation of a nuclear receptor gene, NR2E3, causes enhanced S cone syndrome, a disorder of retinal cell fate. Nat Genet 2000;24:127–131.

23 Feng W, Yasumura D, Matthes MT, LaVail MM, Vollrath D: Mertk triggers uptake of photoreceptor outer segments during phagocytosis by cultured retinal pigment epithelial cells. J Biol Chem 2002;22:22.
24 Izaddoost S, Nam SC, Bhat MA, Bellen HJ, Choi KW: *Drosophila* crumbs is a positional cue in photoreceptor adherens junctions and rhabdomeres. Nature 2002;416:178–183.
25 Den Hollander AI, ten Brink JB, de Kok YJ, van Soest S, van den Born LI, van Driel MA, van de Pol DJ, Payne AM, Bhattacharya SS, Kellner U, Hoyng CB, Westerveld A, Brunner HG, Bleeker-Wagemakers EM, Deutman AF, Heckenlively JR, Cremers FP, Bergen AA: Mutations in a human homologue of *Drosophila* crumbs cause retinitis pigmentosa (RP12). Nat Genet 1999;23:217–221.
26 Lotery AJ, Jacobson SG, Fishman GA, Weleber RG, Fulton AB, Namperumalsamy P, Heon E, Levin AV, Grover S, Rosenow JR, Kopp KK, Sheffield VC, Stone EM: Mutations in the CRB1 gene cause Leber congenital amaurosis. Arch Ophthalmol 2001;119:415–420.
27 Rivolta C, Sweklo EA, Berson EL, Dryja TP: Missense mutation in the USH2A gene: Association with recessive retinitis pigmentosa without hearing loss. Am J Hum Genet 2000;66:1975–1978.
28 Alagramam KN, Yuan H, Kuehn MH, Murcia CL, Wayne S, Srisailpathy CR, Lowry RB, Knaus R, Van Laer L, Bernier FP, Schwartz S, Lee C, Morton CC, Mullins RF, Ramesh A, Van Camp G, Hagemen GS, Woychik RP, Smith RJ: Mutations in the novel protocadherin PCDH15 cause Usher syndrome type 1F. Hum Mol Genet 2001;10:1709–1718.
29 Bork JM, Peters LM, Riazuddin S, Bernstein SL, Ahmed ZM, Ness SL, Polomeno R, Ramesh A, Schloss M, Srisailpathy CR, Wayne S, Bellman S, Desmukh D, Ahmed Z, Khan SN, Kaloustian VM, Li XC, Lalwani A, Bitner-Glindzicz M, Nance WE, Liu XZ, Wistow G, Smith RJ, Griffith AJ, Wilcox ER, Friedman TB, Morell RJ: Usher syndrome 1D and nonsyndromic autosomal recessive deafness DFNB12 are caused by allelic mutations of the novel cadherin-like gene CDH23. Am J Hum Genet 2001;68:26–37.
30 McKie AB, McHale JC, Keen TJ, Tarttelin EE, Goliath R, van Lith-Verhoeven JJ, Greenberg J, Ramesar RS, Hoyng CB, Cremers FP, Mackey DA, Bhattacharya SS, Bird AC, Markham AF, Inglehearn CF: Mutations in the pre-mRNA splicing factor gene PRPC8 in autosomal dominant retinitis pigmentosa (RP13). Hum Mol Genet 2001;10:1555–1562.
31 Vithana EN, Abu-Safieh L, Allen MJ, Carey A, Papaioannou M, Chakarova C, Al-Maghtheh M, Ebenezer ND, Willis C, Moore AT, Bird AC, Hunt DM, Bhattacharya SS: A human homolog of yeast pre-mRNA splicing gene, PRP31, underlies autosomal dominant retinitis pigmentosa on chromosome 19q13.4 (RP11). Mol Cell 2001;8:375–381.
32 Chakarova CF, Hims MM, Bolz H, Abu-Safieh L, Patel RJ, Papaioannou MG, Inglehearn CF, Keen TJ, Willis C, Moore AT, Rosenberg T, Webster AR, Bird AC, Gal A, Hunt D, Vithana EN, Bhattacharya SS: Mutations in HPRP3, a third member of pre-mRNA splicing factor genes, implicated in autosomal dominant retinitis pigmentosa. Hum Mol Genet 2002;11:87–92.
33 Kramer A: The structure and function of proteins involved in mammalian pre-mRNA splicing. Annu Rev Biochem 1996;65:367–409.
34 Hagstrom SA, Adamian M, Scimeca M, Pawlyk BS, Yue G, Li T: A role for the Tubby-like protein 1 in rhodopsin transport. Invest Ophthalmol Vis Sci 2001;42:1955–1962.
35 Roepman R, Bernoud-Hubac N, Schick DE, Maugeri A, Berger W, Ropers HH, Cremers FP, Ferreira PA: The retinitis pigmentosa GTPase regulator (RPGR) interacts with novel transport-like proteins in the outer segments of rod photoreceptors. Hum Mol Genet 2000;9:2095–2105.
36 Bowne SJ, Daiger SP, Hims MM, Sohocki MM, Malone KA, McKie AB, Heckenlively JR, Birch DG, Inglehearn CF, Bhattacharya SS, Bird A, Sullivan LS: Mutations in the RP1 gene causing autosomal dominant retinitis pigmentosa. Hum Mol Genet 1999;8:2121–2128.
37 Liu Q, Zhou J, Daiger SP, Farber DB, Heckenlively JR, Smith JE, Sullivan LS, Zuo J, Milam AH, Pierce EA: Identification and subcellular localization of the RP1 protein in human and mouse photoreceptors. Invest Ophthalmol Vis Sci 2002;43:22–32.
38 Schwahn U, Lenzner S, Dong J, Feil S, Hinzmann B, van Duijnhoven G, Kirschner R, Hemberger M, Bergen AA, Rosenberg T, Pinckers AJ, Fundele R, Rosenthal A, Cremers FP, Ropers HH, Berger W: Positional cloning of the gene for X-linked retinitis pigmentosa 2. Nat Genet 1998;19:327–332.

39 Tubb BE, Bardien-Kruger S, Kashork CD, Shaffer LG, Ramagli LS, Xu J, Siciliano MJ, Bryan J: Characterization of human retinal fascin gene (FSCN2) at 17q25: Close physical linkage of fascin and cytoplasmic actin genes. Genomics 2000;65:146–156.
40 Bowne SJ, Sullivan LS, Blanton SH, Cepko CL, Blackshaw S, Birch DG, Hughbanks-Wheaton D, Heckenlively JR, Daiger SP: Mutations in the inosine monophosphate dehydrogenase 1 gene (IMPDH1) cause the RP10 form of autosomal dominant retinitis pigmentosa. Hum Mol Genet 2002;11:559–568.
41 Kennan A, Aherne A, Palfi A, Humphries M, McKee A, Stitt A, Simpson DA, Demtroder K, Orntoft T, Ayuso C, Kenna PF, Farrar GJ, Humphries P: Identification of an IMPDH1 mutation in autosomal dominant retinitis pigmentosa (RP10) revealed following comparative microarray analysis of transcripts derived from retinas of wild-type and Rho(–/–) mice. Hum Mol Genet 2002;11:547–557.
42 Keen TJ, Hims MM, McKie AB, Moore AT, Doran RM, Mackey DA, Mansfield DC, Mueller RF, Bhattacharya SS, Bird AC, Markham AF, Inglehearn CF: Mutations in a protein target of the Pim-1 kinase associated with the RP9 form of autosomal dominant retinitis pigmentosa. Eur J Hum Genet 2002;10:245–249.
43 Chang GQ, Hao Y, Wong F: Apoptosis: final common pathway of photoreceptor death in rd, rds, and rhodopsin mutant mice. Neuron 1993;11:595–605.
44 Portera-Cailliau C, Sung CH, Nathans J, Adler R: Apoptotic photoreceptor cell death in mouse models of retinitis pigmentosa. Proc Natl Acad Sci USA 1994;91:974–978.
45 Travis GH: Mechanisms of cell death in the inherited retinal degenerations. Am J Hum Genet 1998;62:503–508.
46 Pierce EA: Pathways to photoreceptor cell death in inherited retinal degenerations. Bioessays 2001;23:605–618.
47 Lisman J, Fain G: Support for the equivalent light hypothesis for RP. Nat Med 1995;1:1254–1255.
48 OMIM: Online Mendelian Inheritance in Man *(www3.ncbi.nlm.nih.gov/Omin)*.
49 Steinberg RH: Monitoring communications between photoreceptors and pigment epithelial cells: Effects of 'mild' systemic hypoxia. Friedenwald Lecture. Invest Ophthalmol Vis Sci 1987; 28:1888–1904.
50 Luthert PJ, Chong NH: Photoreceptor rescue. Eye 1998;12:591–596.
51 Sohocki MM, Daiger SP, Bowne SJ, Rodriquez JA, Northrup H, Heckenlively JR, Birch DG, Mintz-Hittner H, Ruiz RS, Lewis RA, Saperstein DA, Sullivan LS: Prevalence of mutations causing retinitis pigmentosa and other inherited retinopathies. Hum Mutat 2001;17:42–51.
52 Vervoort R, Lennon A, Bird AC, Tulloch B, Axton R, Miano MG, Meindl A, Meitinger T, Ciccodicola A, Wright AF: Mutational hot spot within a new RPGR exon in X-linked retinitis pigmentosa. Nat Genet 2000;25:462–466.
53 Rozet JM, Perrault I, Gigarel N, Souied E, Ghazi I, Gerber S, Dufier JL, Munnich A, Kaplan J: Dominant X-linked retinitis pigmentosa is frequently accounted for by truncating mutations in exon ORF15 of the RPGR gene. J Med Genet 2002;39:284–285.
54 Nakazawa M, Wada Y, Tamai M: Arrestin gene mutations in autosomal recessive retinitis pigmentosa. Arch Ophthalmol 1998;116:498–501.
55 Dryja TP, Rucinski DE, Chen SH, Berson EL: Frequency of mutations in the gene encoding the α subunit of rod cGMP-phosphodiesterase in autosomal recessive retinitis pigmentosa. Invest Ophthalmol Vis Sci 1999;40:1859–1865.
56 Acland GM, Aguirre GD, Ray J, Zhang Q, Aleman TS, Cideciyan AV, Pearce-Kelling SE, Anand V, Zeng Y, Maguire AM, Jacobson SG, Hauswirth WW, Bennett J: Gene therapy restores vision in a canine model of childhood blindness. Nat Genet 2001;28:92–95.
57 O'Neill B, Millington-Ward S, O'Reilly M, Tuohy G, Kiang AS, Kenna PF, Humphries P, Farrar GJ: Ribozyme-based therapeutic approaches for autosomal dominant retinitis pigmentosa. Invest Ophthalmol Vis Sci 2000;41:2863–2869.
58 Berson EL: Nutrition and retinal degenerations. Int Ophthalmol Clin 2000;40:93–111.
59 Lund RD, Adamson P, Sauve Y, Keegan DJ, Girman SV, Wang S, Winton H, Kanuga N, Kwan AS, Beauchene L, Zerbib A, Hetherington L, Couraud PO, Coffey P, Greenwood J: Subretinal transplantation of genetically modified human cell lines attenuates loss of visual function in dystrophic rats. Proc Natl Acad Sci USA 2001;98:9942–9947.

60 Zrenner E, Stett A, Weiss S, Aramant RB, Guenther E, Kohler K, Miliczek KD, Seiler MJ, Haemmerle H: Can subretinal microphotodiodes successfully replace degenerated photoreceptors? Vision Res 1999;39:2555–2567.

Chris Inglehearn
Molecular Medicine Unit, Clinical Sciences Building,
St James' University Hospital, Leeds LS9 7TJ (UK)
Tel. +44 113 2065698, Fax +44 113 2444475, E-Mail cinglehe@hgmp.mrc.ac.uk

Wissinger B, Kohl S, Langenbeck U (eds): Genetics in Ophthalmology.
Dev Ophthalmol. Basel, Karger, 2003, vol 37, pp 126–140

Bardet-Biedl Syndrome and Usher Syndrome

Rainer Koenig

Institute of Human Genetics, University Hospital,
Frankfurt am Main, Germany

Abstract

Bardet-Biedl syndrome (BBS) and Usher syndrome (USH) are the most prevalent syndromic forms of retinitis pigmentosa (RP), together they make up almost a quarter of the patients with RP. BBS is defined by the association of retinopathy, obesity, hypogonadism, renal dysfunction, postaxial polydactyly and mental retardation. This clinically complex syndrome is genetically heterogeneous with linkage to more than 6 loci, and 4 genes have been cloned so far. Recent molecular data present evidence that, in some instances, the clinical manifestation of BBS requires recessive mutations in 1 of the 6 BBS loci plus one or two additional mutations in a second BBS locus (tri- or tetra-allelic inheritance). USH is characterized by the combination of congenital or early-onset sensorineural deafness, RP, and variable degrees of vestibular dysfunction. Each of the three clinical types is genetically heterogeneous: 7 loci have been mapped for type 1, three loci for type 2, and two loci for type 3. Currently, 6 USH genes (MYO7A, USH1C, CDH23, PCDH15, USH2A, USH3) have been identified. Pathogenetically, mutations of the USH1 genes seem to result in defects of auditory and retinal sensory cells, the USH 2 phenotype is caused by defects of extracellular matrix or cell surface receptor proteins, and USH3 may be due to synaptic disturbances. The considerable contribution of syndromic forms of RP requires interdisciplinary approaches to the clinical and diagnostic management of RP patients.

Introduction

During the last 10 years an enormous amount on molecular data has widened our understanding of heritable ocular disorders. Particularly in the group of retinal degenerations, the already enormous clinical heterogeneity is outshined now by the extreme genetic heterogeneity: more than 50 retinitis pigmentosa (RP) loci have been mapped and all modes of inheritance are observed. Two syndromes with RP, the Bardet-Biedl syndrome (BBS) and the Usher syndrome (USH) deserve special interest because of their comparatively high prevalence and because of their molecular-genetic complexity. BBS, once thought to be a homogeneous autosomal recessive entity, now turns out to be a model for complex inheritance, with both autosomal recessive and presumably tri- or tetra-allelic inheritance. Also USH is clinically (three types) and genetically (12 loci) heterogeneous. Identical USH mutations may lead to different USH types, thus making the standard classification uncertain. Some USH mutations may also lead to isolated deafness or isolated RP, confuting the traditional distinction between syndromic and nonsyndromic deafness or RP.

Bardet-Biedl Syndrome

Clinical Features

BBS is characterized by retinopathy, mental retardation, polydactyly, obesity and hypogonadism, all signs with great inter- and intra-familial variability. The first patients with this clinical entity were described by Bardet in 1920 and Biedl in 1922. Particularly through the work of Amman, it became clear that BBS and Laurence-Moon syndrome (with ataxia and spastic paraplegia but no polydactyly) are different disorders. Beales et al. [1] recently reviewed the diagnostic criteria for BBS.

Eyes: The retinal disorder is primarily a degeneration of the photoreceptors and affects both rods and cones. Retinal dystrophy is found in more than 90% of the reported cases. Distinctive is the fast progression of visual loss in the teenage years and the relative absence of significant pigment and proliferative changes in the fundus until late stages of the disease. Night blindness is recognized at a mean of 8.5 years. Legal blindness is reached at a mean age of 15.5 years, earlier than in other forms of RP. Eye symptoms rarely start after 18 years of age and almost all patients have suffered visual loss before the third decade. Up to 5 years, electroretinograms or visually evoked responses may be normal, but 90% of children will have an attenuated electroretinogram response by 10 years. Other eye symptoms have been cataract, nystagmus, strabismus, optic atrophy, macular degeneration, glaucoma and microphthalmia.

Mental development: The IQ has been found below 70 in 41% of patients, with 9% being severely retarded. Learning difficulties occur in 62% of patients, 50% show a developmental delay. Speech defects are particularly common. Most authors reported normal neuroradiological findings. Seizures occur only rarely.

Growth: Obesity usually develops in infancy and is localized along the trunk and proximal parts of the limbs. Obesity, which occurs in about 88%, may be accentuated by short stature below the 50th centile in about 64% of the patients.

Renal system: Structural or functional anomalies are present in almost all patients with BBS. Structural abnormalities are unilateral renal agenesis, ectopic and dysplastic kidneys, fetal type lobulation, cysts and diverticula, blunting and clubbing of the calyces, vesico-ureteric reflux, bladder instability [2]. Histologically, increases in mesangial matrix, extensive foot-process fusion, interstitial sclerosis and fibrosis, and glomerular basement membrane abnormalities may be found. Sonographic changes resemble those of polycystic kidneys, but may be missed because of obesity. Functional problems may become apparent usually in the first two decades and do not show a strong correlation with the structural defects. First symptoms can be polyuria, polydipsia, or proteinuria.

Hypogenitalism and genital anomalies: Hypogenitalism occurs in almost all male patients. Cryptorchidism was found in 13–25%. Other malformations are hypoplastic scrotum and hypospadia. Males are usually infertile due to primary gonadal failure, however, 2 patients with children have been reported [1]. In females with BBS the genital abnormalities encompass vaginal atresia, ovarian cysts, uterus duplex, hypoplastic uterus, fallopian tubes and ovaries, hydrometrocolpos, hematocolpos, female hypospadias, persistent urogenital sinus [3]. Most patients have normal menarche, menses and normal secondary sexual characteristics. Several females have given birth to healthy children.

Limb abnormalities: The characteristic limb malformation is postaxial polydactyly affecting the hands and the feet. Mild syndactyly is usually seen between toes 2 and 3. Brachydactyly is common.

Other features of BBS are hearing loss, cardiac anomalies, oligodontia, Hirschsprung's disease, ataxia and poor coordination, emotional instability, diabetes mellitus and asthma.

Inheritance: BBS presents as an autosomal recessive trait (OMIM 209900) with a high rate of consanguinity. It is a matter of discussion whether partial manifestations in heterozygotes, like polydactyly or hypertension, are truly more common in the relatives than in the general population [1]. Prevalence varies between 1 in 59,000–160,000 with unusual high frequencies in Newfoundland and among Bedouins. BBS constitutes about 5% of all cases with RP.

BBS Loci, Genes and Proteins

Linkage analyses revealed substantial genetic heterogeneity: Six BBS loci have been identified to date with evidence for at least a further one. BBS was linked in 1993 to chromosome 16q21 in a large imbred Israeli-Arab family (BBS2). A year after, a second locus was mapped to chromosome 11q13 (BBS1) in 29 families from Northern Europe descent and three Hispanic families. The third and fourth locus were identified on 3p12–13 and 15q23, respectively, in two large Bedouin families in 1994 and 1995. The fifth locus was mapped to 2q31 in a single large Newfoundland kindred [4] and recently a sixth locus was mapped to 20p12 [5]. The presence of a seventh, as yet unmapped, was suggested by Beales et al. [6].

BBS1 (OMIM 209901)

The gene underlying BBS1 [7] is the cause of BBS in 40–50%. It consists of 17 exons and spans about 23 kb with an open reading frame of 593 codons. The protein does not show significant similarity to known proteins, but weak similarity to BBS2. The most common cause of BBS (one-third of all patients) is the BBS1 allele Met390Arg.

BBS2 (OMIM 606151)

The BBS2 gene causes BBS in 8–16%. It has an open reading frame of 2,163 bp, distributed over 17 exons, and bears no homology to any known protein and does not contain any recognizable motifs [8]. Computational modeling of the three-dimensional configuration of BBS2 predicts the presence of a coiled-coil domain near the N-terminus of the protein, which may indicate protein-protein interactions [9]. The gene displays a wide pattern of tissue expression, including the brain and kidney [8].

BBS3 (OMIM 600151)

The locus is involved in 2–4% of BBS cases and is mapped within a 2-cM region on 3p13.

BBS4 (OMIM 600374)

The gene BBS4 causing 1–3% of BBS [10, 11] is composed of 16 exons and spans about 52 kb. It is ubiquitously expressed, including fetal tissue, retina, adipose tissue and kidney. BBS4 shows strongest homology to O-linked N-acetylglucosamine (O-GlcNAc) transferase from several species. O-GlcNAcs are known to play an important role in signal transduction [12]. In addition, BBS4 contains a potential tetratricopeptide repeat (TPR) motif. Most TPR-containing proteins are associated with multiprotein complexes, and there is extensive evidence that TPR motifs are important to the functioning of chaperone, cell-cycle, transcription and protein transport complexes [13].

BBS5 (OMIM 603650)

The locus [4] is involved in 3% of BBS cases. It is mapped within a 6-cM region on 2q31.

BBS6 = MKKS (OMIM 604896)

Four to 5% of BBS families are linked to the BBS6 locus. The identification of BBS6 [5, 14, 15] was facilitated by the cloning of the gene for McKusick-Kaufman syndrome (MKKS; OMIM 236700) [14]. Through polydactyly, hydrometrocolpos and heart defects, this syndrome shows phenotypic overlap with BBS. Remarkably, there were several patients with MKKS, who later developed retinal dystrophy and obesity [16, 17]. When it became clear that the BBS6 critical region contains MKKS, mutations in MKKS were identified in patients with typical BBS [5, 15]. The MKKS/BBS6 transcript has six exons, a predicted open reading frame of 570 amino acids and is widely expressed [14]. The protein has homology to the group II class chaperonins (archebacterial chaperonins and eukaryotic T-complex-related proteins) with best similarity to the fold of the thermosome from *Thermoplasma acidophilum*. Modeling of the three-dimensional structure of the MKKS protein showed that the missense mutations in MKKS are sited in the predicted highly conserved equatorial domain of the protein. This domain is responsible for ATP hydrolysis in conjunction with facilitated protein folding [14, 18]. Chaperones are proteins that, beside other biological functions, control and catalyze correct protein folding. By binding exposed hydrophobic patches on proteins, chaperons prevent damaged proteins (resulting from stress or gene mutations) from aggregating in insoluble, nonfunctional inclusions or destruction by cytoplasmic proteases [19]. It was therefore hypothesized that the clinical features of BBS may be caused by disturbed protein integrity in the affected organs [15].

BBS7

This locus [6] is negatively defined so far through families without linkage to any of the 6 BBS loci.

BBS Phenotype-Genotype Correlations

The different BBS loci and genes lead to clinically undistinguishable phenotypes. In the known genes – BBS1, BBS2, BBS4, BBS6 – all kinds of mutations were found: frameshifts, missense, nonsense and splice aberrations [20]. It is speculated that the missense mutations cause structural abnormalities that may result in a functionally null protein [5]. Patients with homozygous or compound heterozygous frameshift mutations apparently show no different phenotype. In BBS6 it has suggested that a total loss would lead to the BBS phenotype, whereas milder (hypomorphic) alleles may lead to MKKS [15].

Tri-Allelic Inheritance

In 2001, Katsanis et al. [20] reported a surprising discovery: patients in 4 out of 163 BBS families did not show two affected alleles, as expected for an autosomal recessive disease, but three affected alleles instead. The authors called this phenomenon tri-allelic inheritance, because one of the six BBS loci has both alleles of the gene mutated, and another one out of the remaining five loci has one mutant allele in addition. Even more impressive is the fact that 47% of their families with BBS2 and 37% with BBS6 mutations show involvement of another locus taking together all genetic data. Multilocus, multiallelic inheritance therefore may not be an exception in BBS. The tri-allelic hypothesis was further supported by the observation that 2 unaffected individuals carried two BBS2 mutations only and were wild-type for BBS6 in contrast to the affected relatives [9, 20]. Mutation screening of BBS4 indicated that in some instances more than three mutant alleles may be required for the manifestation of the BBS phenotype [11]. The tri-allelic hypothesis was recently questioned by Mykytyn et al. [7]. They showed that a common BBS1 missense mutation is a frequent cause of BBS, but that this mutation is not involved in tri-allelic inheritance. This leads the discussion to questions of multi-allelic inheritance or recessive inheritance with modifier or susceptibility genes, questions which are to bridge classical Mendelian and multifactorial inheritance.

Usher Syndrome

Clinical Features

Usher syndrome USH designates a group of clinically and genetically heterogeneous autosomal recessive disorders characterized by the combination of congenital or early-onset sensorineural deafness and RP. The syndrome is the most frequent cause of deafness accompanied by blindness. USH is reported to account for between 3 and 6% of the congenital deaf patients, about 18% of those with RP, and for more than 50% of the deaf-blind patients. Its prevalence is estimated between 1/16,000 and 1/50,000.

The association of congenital deafness and progressive pigmentary retinopathy was first described by the German ophthalmologist Albrecht von Graefe in 1858. In 1914, the British ophthalmologist Charles Usher confirmed the autosomal recessive inheritance and suggested the existence of at least two clinical types of the disease, according to the degree of hearing impairment and the age of onset and progression of visual loss. In 1977, Davenport and Omenn [21] defined three subtypes, USH1–3, which are still in clinical use (table 1).

Clinical criteria recommended for the diagnosis of USH have been defined recently by the Usher Consortium [22].

Eyes: Night blindness, often the first symptom of RP, may appear as early as preschool age. Abnormalities in electroretinogram may precede visual symptoms, which allows diagnosis in early childhood. Visual loss increases due to

Table 1. Usher syndrome – clinical classification and major findings

Type	Hearing loss	Vestibular function	Night blindness	Visual field loss	Frequency
USH1	Congenital, severe	Absent	1st–early 2nd decade	Severe, usually in the 1st decade	33–44%
USH2	Moderate to profound, low frequencies usually preserved	Normal	Late 2nd–early 3rd decade	Variable, normal to severe	56–67%
USH3	Progressive	Mostly normal	Variable	Variable	10–20% (?)

progressive degeneration of rod photoreceptor cells, but it deteriorates slowly, progressing to blindness in about 40% in the fifth decade, about 60% in the sixth decade, and in about 75% in the seventh decade in all types of USH. Ophthalmologic examination shows the typical picture of progressive RP consisting of bony spicule pigment clumping, which begins in the midperiphery of the fundus and extends inward and outward. Later, the optic discs become pale and the arterioles become narrowed.

Auditory and vestibular system: Patients with type I (USH1) were born with severe to profound congenital sensorineural deafness and absent vestibular function. Residual low frequency hearing may be detectable at 90–100 dB, but 'islands' of hearing between 250 and 8,000 Hz hearing do not exist. Therefore, traditional hearing aids are not helpful in these patients, but they may benefit from cochlear implantation. In USH2 the audiogram shows a typical sloping from a moderate hearing loss for the low frequencies down to a severe loss in the high frequencies. Thus, patients with USH2 usually benefit from hearing aids. Progressive hearing loss is typical for USH3 and distinguishes this type from USH1 and USH2. Patients with USH1 show a markedly reduced vestibular response to caloric or rotational testing, whereas testing is normal in type II and variable in type III. The dysfunction of the vestibular system in USH1 may lead to a delay in motor development.

Pathology: Abnormalities of cilial cells, the proposed pathogenetic targets in USH, may explain the involvement of three sensory systems. Photoreceptors, auditory hair cells and vestibular hair cells develop from ciliated progenitors, and axonemes are present in mature photoreceptors and vestibular hair cells.

Histopathological studies show a severe degeneration of the organ of Corti and atrophy of the spiral ganglia in all USH types [23]. Abnormal axonemes occur in remnant photoreceptors [24]. The authors found a destabilization of the photoreceptor, which is indicated first by shortened inner segments and stacked discs in the outer segments. In the next stage, there were no photoreceptor outer segments or connecting cilia, whereas remnant inner segments were present. At last there were no remnant photoreceptors and the pigment epithelial layer was directly apposed to the apical processes of the Müller cells. The more general involvement of ciliae in USH patients is documented by abnormalities of nasal cilia and decreased sperm motility with collapsed axonemes in the sperm tail.

USH Loci, Genes, and Proteins

Autosomal recessive inheritance is well established for USH. The three clinical types exhibit a significant genetic heterogeneity: there are 7 mapped loci for type 1 (USH1A, USH1B, USH1C, USH1D, USH1E, USH1F, USH1G), three loci for type 2 (USH2A, USH2B, USH2C), and two loci for type 3 (USH3A, USH3B). Moreover, families have been found without linkage to any of the known loci, indicating additional USH genes. Actualized information about mapped loci and genes for phenotypes involving hearing loss are presented on *http://www.uia.ac.be/dnalab/hhh/*

USH1A (OMIM 276900)

The locus is on 14q32. No gene has been identified yet.

USH1B (OMIM 276903)

USH1B on 11q13.5 is caused by mutations in the gene MYO7A, which encodes an unconventional type VII myosin. The identification of the gene was facilitated by studies of the recessive mouse-deafness mutant *shaker-1 (sh-1)*, which exhibits circling and hyperactivity and progressive loss of hearing and balance in the first postnatal weeks. MYO7A consists of 48 coding exons and spans over 100 kb. Most MYO7A mutations cause typical USH1 phenotypes, but some cause nonsyndromic autosomal recessive deafness (DFNB2), autosomal dominant deafness (DFNA11), or atypical USH with progressive hearing loss. Re-examination after 7 years of a large family, first classified as nonsyndromic deafness (DFNB2), revealed that several members had developed a mild retinal degeneration in addition to progressive deafness. The authors concluded that environmental and/or genetic factors likely modulate the phenotypic effect of certain myosin VIIa mutations, leading to inter- and intra-familial variability of both the hearing loss and the retinal dystrophy [25].

Unconventional myosins [26] are motor proteins, which move along actin filaments using ATP hydrolysis to produce energy. In the inner ear, myosin VIIa has been localized in the sensory hair cells of both auditory and vestibular sensory epithelia. It is mostly found in a band near the ankle links and in the pericuticular necklace. The first place points to a stabilizing function of myosin VIIa, which may hold the stereocilia in register. The second place is thought to be a region of membrane trafficking, suggesting that myosin VIIa may be bound to vesicles and may be part of a transport mechanism. In the retina, myosin VIIa is found in the pigment epithelium cells and the photoreceptor cells. Rhodopsin is translocated from the inner segment to the outer segment of the photoreceptor via the ciliary membrane, at which rhodopsin co-localizes with myosin VIIa [27]. In addition, axonemal actin is spatially co-localized in the photoreceptor cilium with myosin VIIa and opsin [27]. Thus, the actin skeleton of the cilium may provide the structural basis for myosin VIIa-linked ciliary trafficking of membrane components, including rhodopsin.

USH1C (OMIM 276904, 605242)

USH1C on 11p15.1 encodes a PDZ domain-containing protein, named harmonin [28, 29]. It consists of 28 coding exons and spans about 51 kb. So far, six mutations have been described, all predicted to cause a truncated protein. An expanded intronic VNTR was described in an affected Acadian family, possibly resulting in interruption of transcriptional or posttranscriptional processing [28]. The 238–239insC mutation of several European patients points to a founder effect. Alternative splicing results in a great variety of transcripts, which differ, among other things, in the number of their PDZ- or coiled-coil domains. Some isoforms are expressed either in the hair cells of the inner ear [28] or in the photoreceptor [29]. PDZ domains derive their name from the first proteins recognized to have the common conserved motif of 80–90 amino acids: the postsynaptic density protein *P*SD95, the *Drosophila melanogaster* tumor suppressor gene *D*lgA and the tight junction protein *Z*O-1. It is suggested that PDZ proteins localize their ligands (receptors, channels, components of signal transduction) to specific subcellular compartments and organize and coordinate multiprotein complexes at the plasma membrane [29]. In this regard they may bridge or modulate signaling pathways to the cytoskeleton.

USH1D (OMIM 601067, 605516)

The USH1D locus on 10q22 was defined in a consanguineous Pakistani family by homozygosity mapping. A novel cadherin-like gene, CDH23, was identified 5 years later [30]. The gene consists of 69 coding exons, spans more than 300 kb and is highly expressed in the retina and in the hair cells of the ear. Cadherins and cadherin-like proteins mediate cell-cell or cell-extracellular membrane interactions. CDH23 is a single-pass transmembrane protein with 27 extracellular cadherin repeats and a cytoplasmic region without homology to known proteins. Recessive *waltzer* (*v*) mice, the orthologous mouse models of USH1D, show disorganized stereocilia, suggesting that CDH23 may function as a 'hair bundle organizer' [31]. As in the case of the *shaker-1* mutants, *waltzer* mice show no gross abnormalities of the retina. Interestingly, protein truncating mutations in humans lead to the USH1 phenotype, whereas missense mutations result in nonsyndromal deafness (DFNB12) or atypical USH with only mild retinal involvement [32].

USH1E (OMIM 602097)

In this locus on 21q no causative gene has been identified yet.

USH1F (OMIM 602083, 605514)

The USH1F locus was mapped to chromosome 10q11.2. The causative USH1F gene codes for protocadherin 15 (PCDH15) [33, 34]. The mouse ortholog is *Ames waltzer* (*av*). DFNB23 maps to the same locus. The PCDH15 gene consists of 33 exons and spans about 1.6 Mb. It is predicted to have 11 extracellular calcium-binding domains, a transmembrane domain and a unique intracellular domain. Protocadherins are thought to be involved in a variety of functions, including neural development and formation of the synapse. The growth and arrangement of stereocilia in the hair cells to the typical V-shaped bundles is a process of planar polarity, which may be regulated by PCDH15 [33].

USH1G (OMIM 606943)

USH1G was mapped to chromosome 17q24–25 [35]. This region overlaps with the intervals for the two dominant nonsyndromic forms of deafness, DFNA20 and DFNA26. The locus of the recessive mouse mutant *Jackson shaker* (*js*) resides in a segment of mouse chromosome 11, that is homologous with human chromosome 17q25. Similar to the other mouse models, *js* mice do not have retinal degeneration.

USH2A (OMIM 276901)

USH2A was the first Usher locus mapped. The gene on 1q41 encodes a protein, termed usherin, which resembles extracellular matrix and cell surface receptor proteins [36]. It contains a laminin domain 6-like (L6) motif, 10 laminin-type EGF-like (LE), and 4 fibronectin-like type 3 (F3) domains. The signal peptide following region and the C-terminal region have no homology to known proteins. Laminin-6 motifs are found, beside others, in netrins, which are involved in neural-glial interactions, like axon guidance. Laminins are the main noncollagenous component of the basement membranes. Usherin was found in the basement membranes of the inner ear and retina.

However, in contrast to the USH1 proteins, it was not observed in the hair cells of the cochlear and in the photoreceptor cells [37]. Therefore, it is likely that usherin affects other pathogenetic pathways than the USH1 proteins. More than 20 mutations have been identified, most of which are nonsense, splice, missense and frameshift mutations. The accumulation of missense mutations in the laminin type 6 domain point to a critical functional role of this region. There is one mutational hotspot (2299delG), which is detected in about 20% of USH2A families in Europe and in the USA [38]. Similar to USH1 mutations, USH2A mutations may lead to atypical phenotypes. Even monozygotic (male) twins with the same 2299delG mutation were found discordant, presenting with USH2 and USH3 phenotypes, respectively [39]. This indicates that variations in the genetic background are not the only causes of phenotypic variability, but that stochastic events may also play a role. The missense mutation Cys759Phe in the fifth LE motif of the USH2A gene is remarkable, because it is detected in 4.5% of patients with nonsyndromic RP [40], and because all other allelic forms of USH lead to nonsyndromic deafness.

USH2B (OMIM 276905)

In this locus on 3p23–p24.2 [41] no causative USH gene has yet been found. The region overlaps with the nonsyndromic deafness locus DFNB6.

USH2C (OMIM 605472)

This locus is on 5q14.3–q21.3 [42].

USH3A (OMIM 276902, 606397)

Type 3 USH was mapped to 3q21–q25 in Finish families. Through linkage disequilibrium and mutation analysis, the USH3A gene was identified [43]. It codes for the protein Clarin-1 [44]. The ORF predicts a 232-amino-acid protein with four transmembrane domains (4TM) and a single glycosylation consensus site. All mutations, found so far, are thought to result in an inactive protein. Based on its sequence similarities to stargazin, Clarin-1 may play a role in hair cell and photoreceptor cell synapses [44].

Another USH3 locus has been suggested for 20q. This assignment is not yet validated.

USH Phenotype-Genotype Correlations

The clinical differentiation between USH1, USH2 and USH3 apparently correlates with different pathogenetic pathways. Abnormalities of cilial cells are typical for USH1, whereas USH2 results from abnormalities of the basement membranes and USH3 is possibly caused by synaptic disturbances.

Protein truncating mutations of CDH23 produce the severe USH1 phenotype, whereas missense mutations result in nonsyndromal deafness or atypical, less severe USH. However, the genotype does not always predict the phenotype, documented by variability within and between families carrying the same mutations and even discordant monocygotic twins.

Conclusions

Recent investigations in BBS and USH have provided significant advances both in the genetics and in the pathophysiology. Apparently, two alleles are not always sufficient to manifest BBS, but three or more. This new type of multiallelic, multiloci inheritance must be further substantiated. As the 4 known BBS genes also do not result in clinically distinguishable phenotypes, it may be argued that their products interact directly or affect the same pathway. Whether and how these proteins interact will depend on elucidation of their biochemistry.

The identification of the 6 USH genes has confirmed the clinical classifications in various subtypes, which probably also reflect different pathogenetic pathways. Moreover, mutations of the same gene can lead to phenotypes from isolated deafness to severe USH1 syndrome. Therefore, the traditional distinction between syndromic and nonsyndromic deafness/RP is no longer strict, but becomes vague. Careful, repeated and perhaps more sophisticated clinical and neurophysiological examinations are necessary to permit a more accurate genotype-phenotype correlation. A direct clinical consequence may be visual/auditory screening of patients with so-called isolated nonsyndromic deafness or RP.

References

Due to limitations of space, only a selection of references has been given. For full lists of citations see the knowledgebase OMIM (Online Mendelian Inheritance in Man) at *http://www.ncbi.nlm.nih.gov/Omim/*

1 Beales PL, Elcioglu N, Woolf AS, Parker D, Flinter FA: New criteria for improved diagnosis of Bardet-Biedl syndrome: Results of a population survey. J Med Genet 1999;36:437–446.

2 O'Dea D, Parfrey PS, Harnett D, Hefferton D, Cramer BC, Green J: The importance of renal impairment in the natural history of Bardet-Biedl syndrome. Am J Kidney Dis 1996;27:776–783.

3 Stoler JM, Herrin JT, Holmes LB: Genital abnormalities in females with Bardet-Biedl syndrome. Am J Med Genet 1995;55:276–278.

4 Young TL, Penney L, Woods MO, Lewis RA, Green JS, Hefferton D, Davidson WS: A fifth locus for Bardet-Biedl syndrome maps to 2q31. Am J Hum Genet 1999;64:901–904.

5 Katsanis N, Beales PL, Woods MO, Lewis RA, Green JS, Parfrey PS, Ansley SJ, Davidson WS, Lupski JR: Mutations in MKKS cause obesity, retinal dystrophy and renal malformations associated with Bardet-Biedl syndrome. Nat Genet 2000;26:67–70.

6 Beales PL, Katsanis N, Lewis RA, Ansley SJ, Elcioglu N, Raza J, Woods MO, Green JS, Parfrey PS, Davidson WS, Lupski JR: Genetic and mutational analyses of a large multiethnic Bardet-Biedl

cohort reveal a minor involvement of BBS6 and delineate the critical intervals of other loci. Am J Hum Genet 2001;68:606–616.
7 Mykytyn K, Nishimura DY, Searby CC, Shastri M, Yen HJ, Beck JS, Braun T, Streb LM, Cornier AS, Cox GF, Fulton AB, Carmi R, Lüleci G, Chandrasekharappa SC, Collins FC, Jacobson SG, Heckenlively JR, Weleber RG, Stone EM, Sheffield VC: Identification of the gene (BBS1) most commonly involved in Bardet-Biedl syndrome, a complex human obesity syndrome. Nat Genet 2002;31:435–438.
8 Nishimura DY, Searby CC, Carni R, Elbedour K, Van Maldergem L, Fulton AB, Lam BL, Powell BR, Swiderski RE, Bugge KE, Haider NB, Kwiteck-Black AE, Ying L, Duhl DM, Gorman SW, Heon E, Iannaccone A, Bonneau D, Biesecker LG, Jacobson SG, Stone EM, Sheffield VC: Positional cloning of a novel gene on chromosome 16q causing Bardet-Biedl syndrome (BBS2). Hum Mol Genet 2001;10:865–874.
9 Katsanis N, Lupski JR, Beales PL: Exploring the molecular basis of Bardet-Biedl syndrome. Hum Mol Genet 2001;10:2293–2299.
10 Mykytyn K, Braun T, Carmi R, Haider NB, Searby CC, Shastri M, Beck G, Wright AF, Iannaccone A, Elbedour K, Riise R, Baldi A, Raas-Rothschild A, Gorman SW, Duhl DM, Jacobson SG, Casavant T, Stone EM, Sheffield VC: Identification of the gene that, when mutated, causes the human obesity syndrome BBS4. Nat Genet 2001;28:188–191.
11 Katsanis N, Eichers ER, Ansley SJ, Lewis RA, Kayserili H, Hoskins BE, Scambler PJ, Beales PL, Lupski JR: BBS4 is a minor contributor to Bardet-Biedl syndrome and may also participate in triallelic inheritance. Am J Hum Genet 2002;71:22–29.
12 Wells L, Vosseller K, Hart GW: Glycosylation of nucleocytoplasmic proteins: Signal transduction and O-GlcNAc. Science 2001;291;2376–2378.
13 Blatch GL, Lässle M: The tetratricopeptide repeat: A structural motif mediating protein-protein interactions. Bioessays 1999;21:932–939.
14 Stone DL, Slavotinek A, Bouffard GG, Banerjee-Basu S, Baxevanis AD, Barr M, Biesecker LG: Mutations in a gene encoding a putative chaperonine causes McKusick-Kaufman syndrome. Nat Genet 2000;25:79–82.
15 Slavotinek AM, Stone EM, Mykytyn K, Heckenlively JR, Green JS, Heon E, Musarella MA, Parfrey PS, Sheffield VC, Biesecker LG: Mutations in MKKS cause Bardet-Biedl syndrome. Nat Genet 2000;26:15–16.
16 Shaap C, ten Tusscher MP, Schrander JJ, Kuijten RH, Schrander-Stumpel CT: Phenotypic overlap between McKusick-Kaufman and Bardet-Biedl syndromes: Are they related? Eur J Pediatr 1998;157:170–171.
17 David A, Bitoun P, Lacombe D, Lambert JC, Nivelon A, Vigneron J, Verloes A: Hydrometrocolpos and polydactyly: A common neonatal presentation of Bardet-Biedl and McKusick-Kaufman syndromes. J Med Genet 1999;36:599–603.
18 Slavotinek AM, Biesecker LG: Unfolding the role of chaperones and chaperonins in human disease. Trends Genet 2001;17:528–535.
19 Wickner S, Maurizi MR, Gottesman S: Posttranslational quality control: Folding, refolding and degrading proteins. Science 1999;286:1888–1893.
20 Katsanis N, Ansley SJ, Badano JL, Eichers ER, Lewis RA, Hoskins BE, Scambler PJ, Davidson WS, Beales PL, Lupski JR: Triallelic inheritance in Bardet-Biedl syndrome, a Mendelian recessive disorder. Science 2001;293:2256–2259.
21 Davenport SLH, Omenn GS: The heterogeneity of Usher syndrome. Int Congr Ser, Publ 426. Amsterdam, Excerpta Medica Foundation, 1977, abstr 215, pp 87–88.
22 Smith RJH, Berlin CI, Hejtmancik JF, Keats BJB, Kimberling WJ, Lewis RA, Möller CG, Pelias MT, Tranebjaerg L: Clinical diagnosis of Usher sydromes. Am J Med Genet 1994;50:32–38.
23 Wagenaar M, Schuknecht H, Nadol J Jr, Benraad-Van Rens M, Pieke-Dahl S, Kimberling W, Cremers C: Histopathologic features of the temporal bone in Usher syndrome type I. Arch Otolaryngol Head Neck Surg 2000;126:1018–1023.
24 Berson EL, Adamian M: Ultrastructural findings in an autopsy eye from a patient with Usher's syndrome type II. Am J Ophthalmol 1992;114:748–757.
25 Zina ZB, Masmoudi S, Ayadi H, Chaker F, Ghorbel AM, Drira M, Petit C: From DFNB2 to Usher syndrome: Variable expressivity of the same disease. Am J Med Genet 2001;101:181–183.

26 Friedman TB, Sellers JR, Avraham KB: Unconventional myosins and the genetics of hearing loss. Am J Med Genet 1999;89:147–157.
27 Wolfrum U, Schmidt A: Rhodopsin transport in the membrane of the connecting cilium of mammalian photoreceptor cells. Cell Motil Cytoskeleton 2000;46:95–107.
28 Verpy E, Leibovici M, Zwaenepoel I, Xue-Zhong L, Gal A, Salem N, Mansour A, Blanchard S, Kobayashi I, Keats BJB, Slim R, Petit C: A defect in harmonin, a PDZ domain-containing protein expressed in the inner ear sensory hair cells, underlies Usher syndrome type 1 C. Nat Genet 2000;26:51–55.
29 Bitner-Glindizicz M, Lindley KJ, Rutland P, Blaydon D, Smith VV, Milla PJ, Hussain K, Furth-Lavi J, Cosgrove KE, Shepherd RM, Barnes PD, O'Brien RE, Farndon PA, Sowden J, Liu XZ, Scanlan MJ, Malcolm S, Dunne MJ, Aynsley-Green A, Glaser B: A recessive contiguous gene deletion causing infantile hyperinsulinism, enteropathy and deafness identifies the Usher type 1C gene. Nat Genet 2000;26:56–60.
30 Bolz H, von Brederlow B, Ramirez A, Bryda EC, Kutsche K, Nothwang HG, Seeliger M, Cabrera M, Vila MC, Molina OP, Gal A, Kubisch C: Mutation of CDH23, encoding a new member of the cadherin gene family, causes Usher syndrome type 1D. Nat Genet 2001;27:108–112.
31 Di Palma F, Holme RH, Bryda E, Belyantseva IA, Pellegrino R, Kachar B, Steel KP, Noben-Trauth K: Mutations in *Cdh23*, encoding a new type of cadherin, cause stereocilia disorganization in waltzer, the mouse model for Usher syndrome type 1D. Nat Genet 2001;27:103–107.
32 Bork JM, Peters LM, Riazuddin S, Bernstein SL, Ahmed ZM, Ness SL, Polomeno R, Ramesh A, Schloss M, Srisailpathy CRS, Wayne S, Bellman S, Desmukh D, Ahmed Z, Khan SN, der Kaloustian VM, Li XC, Lalwani A, Riazuddin S, Bitner-Glindzicz M, Nance WE, Liu XZ, Wistow G, Smith RJH, Griffith AJ, Wilcox ER, Friedman TB, Morell RJ: Usher syndrome 1D and non-syndromic autosomal recessive deafness DFNB12 are caused by allelic mutations of the novel cadherin-like gene CDH23. Am J Hum Genet 2001;68:26–37.
33 Alagramam KN, Yuan H, Kuehn MH, Murcia CL, Wayne S, Srisailpathy CR, Lowry RB, Knaus R, Van Laer L, Bernier FP, Schwartz S, Lee C, Morton CC, Mullins RF, Ramesh A, Van Camp G, Woychik RP, Smith RJ, Hageman GS: Mutations in the novel protocadherin PCDH15 cause Usher syndrome type 1F. Hum Mol Genet 2001;10:1709–1718.
34 Ahmed ZM, Riazuddin S, Bernstein SL, Ahmed Z, Khan S, Griffith AJ, Morell RJ, Friedman TB, Riazuddin S, Wilcox ER: Mutations of the protocadherin gene PCDH15 cause Usher syndrome type 1F. Am J Hum Genet 2001;69:25–34.
35 Mustapha M, Chouery E, Torchard-Pagnez D, Nouaille S, Khrais A, Sayegh FN, Mégarbané A, Loiselet J, Lathrop M, Petit C, Weil D: A novel locus for Usher syndrome type I, USH1G, maps to chromosome 17q24–25. Hum Genet 2002;110:348–350.
36 Eudy JD, Weston MD, SuFang Y, Hoover DM, Rehm HL, Manling ME, Yan D, Ahmad I, Cheng JJ, Ayuso C, Cremers C, Davenport S, Moller C, Talmadge CB, Beisel KW, Tamayo M, Morton SC, Swaroop A, Kimberling WJ, Sumegi J: Mutation of a gene encoding a protein with extra-cellular matrix motifs in Usher syndrome type IIa. Science 1998;280:1753–1757.
37 Bhattacharya G, Miller C, Kimberling WJ, Jablonski MM, Cosgrove D: Localization and expression of usherin: A novel basement membrane protein defective in people with Usher's syndrome type IIa. Hear Res 2002;163:1–11.
38 Weston MD, Eudy JD, Fujita S, Yao S, Usami S, Cremers C, Greenberg J, Ramesar R, Martina A, Moller C, Smith RJ, Sumegi J, Kimberling WJ, Greenberg J: Genomic structure and identification of novel mutations in usherin, the gene responsible for Usher syndrome type IIa. Am J Hum Genet 2000;66:1199–1210.
39 Liu XZ, Hope C, Liang CY, Zou JM, Xu LR, Cole T, Mueller RF, Bundey S, Nance W, Steel KP, Brown SDM: A mutation (2314delG) in the Usher syndrome type IIA gene: High prevalence and phenotypic variation. Am J Hum Genet 1999;64:1221–1225.
40 Rivolta S, Swelko EA, Berson EL, Dryja TP: Missense mutation in the USH2A gene: Association with recessive retinitis pigmentosa without hearing loss. Am J Hum Genet 2000;66: 1975–1978.
41 Hmani M, Ghorbel A, Boulila-Elgaied A, Zina ZB, Kammoun W, Drira M, Chaabouni M, Petit C, Ayadi H: A novel locus for Usher syndrome type II, USH2B, maps to chromosome 3 at p23–24. Eur J Hum Genet 1999;7:363–367.

42 Pieke-Dahl S, Möller CG, Kelley PM, Astuto LM, Cremers CWRJ, Gorin MB, Kimberling WJ: Genetic heterogeneity of Usher syndrome type II: Localisation to chromosome 5q. J Med Genet 2000;37:256–262.
43 Joensuu T, Hämäläinen R, Yuan B, Johnson C, Tegelberg S, Gasparini P, Zelante L, Privola U, Pakarinen L, Lehesjoki AE, de la Chapelle A, Sankila EM: Mutations in novel gene with transmembrane domains underlie Usher syndrome type 3. Am J Hum Genet 2001;69:673–684.
44 Adato A, Vreugde S, Joensuu T, Avidan N, Hamalainen R, Belenkiy O, Olender T, Bonne-Tamir B, Ben-Asher E, Espinos C, Millán JM, Lehesjoki AE, Flannery JG, Avraham KB, Pietrovski S, Sankila EM, Beckmann JS, Lancet D: USH3A transcripts encode clarin-1, a four-transmembrane-domain protein with a possible role in sensory synapses. Eur J Hum Genet 2002;10:339–350.

Priv.-Doz. Dr. Rainer Koenig
Institute of Human Genetics, University Hospital,
Theodor-Stern-Kai 7, D–60590 Frankfurt/Main (Germany)
Tel. +49 69 6301 6416, Fax +49 69 6301 6002, E-Mail r.koenig@em.uni-frankfurt.de

Wissinger B, Kohl S, Langenbeck U (eds): Genetics in Ophthalmology.
Dev Ophthalmol. Basel, Karger, 2003, vol 37, pp 141–154

Genetic Defects in Vitamin A Metabolism of the Retinal Pigment Epithelium

Debra A. Thompson[a,b], *Andreas Gal*[c]

Departments of [a]Ophthalmology and Visual Sciences, and [b]Biological Chemistry, University of Michigan Medical School, Ann Arbor, Mich., USA and [c]Institut für Humangenetik, Universitätsklinikum Hamburg-Eppendorf, Hamburg, Deutschland

Abstract

The metabolism of vitamin A and cycling of retinoids between the retinal pigment epithelium (RPE) and the neural retina is a complex process involving a specialized enzymes and proteins. Mutations in a number of the corresponding genes are responsible for various forms of inherited retinal dystrophy and dysfunction. Research into the causes and treatment of retinal diseases resulting from defects in vitamin A metabolism is currently the subject of intense interest, since disorders affecting RPE function are, in principle, more accessible to therapeutic intervention than those affecting the proteins of the photoreceptor cells. In this chapter we present an overview of the visual cycle, as well as the function of the known RPE genes involved in the conversion of vitamin A (all-*trans* retinol) to 11-*cis* retinal, the chromophore of the visual pigments. We describe the identification of disease-associated mutations in this set of genes in patients with diverse forms of retinal dystrophy and dysfunction, as well as the spectrum of mutations and associated phenotypes. We also discuss the results of recent studies using animal models of the disease caused by mutations of RPE65. On the basis of these advances, it is hoped that patients with defects in RPE vitamin A metabolism will be among the first successfully treated by targeted therapies likely to become available in the near future.

Introduction

The visual process (phototransduction) begins when light is absorbed by visual pigments, rhodopsin and cone opsins, in the photoreceptor cells. Single photon capture results in the isomerization of the light-sensitive chromophore,

11-*cis* retinal, to all-*trans* retinal and the formation of a protein-excited state [1]. Sustained phototransduction depends on cycling of vitamin A analogs (the visual cycle) between the photoreceptor cells and the retinal pigment epithelium (RPE), the site of 11-*cis* retinal synthesis [2]. All-*trans* retinal is released from the visual pigments following decay of the protein-excited state through a series of thermal intermediates. In rod cells, the released aldehyde can react with phosphatidylethanolamine in the disc membrane to form N-retinylidene-phosphatidylethanolamine that is transferred across the membrane by the retina-specific ATP-binding cassette transporter (ABCR) [3, 4]. All-*trans* retinal is then released into the cytoplasm and reduced to all-*trans* retinol by a short chain acyl-CoA dehydrogenase/reductase and transported to the RPE most likely associated with interstitial retinol-binding protein [reviewed in 5]. All-*trans* retinol also enters the RPE via the choroidal vasculature in a likely receptor-mediated process involving the recognition of a complex with serum retinol-binding protein/transthyretin. Within the RPE, all-*trans* retinol is bound to cellular retinol-binding protein.

The conversion of vitamin A to 11-*cis* retinal requires at least three enzyme activities associated with the RPE smooth endoplasmic reticulum, lecithin retinyl acyltransferase (LRAT), retinoid isomerase, and 11-*cis* retinol dehydrogenase (11 cisRDH), as well as an RPE-specific protein, RPE65 [reviewed in 6]. The retinoid isomerase is also referred to as the isomerohydrolase, as the isomerization reaction has been proposed to be energetically linked to the release of 11-*cis* retinol from an ester intermediate in the membrane bilayer [7]. Recent studies of the genetic defects responsible for inherited photoreceptor degeneration identified mutations in a number of genes that encode proteins involved in RPE vitamin A metabolism. This chapter presents a summary of this recent literature. Genes are discussed in the order in which their disease associated mutations were first discovered. Mutations in genes encoding proteins that interact with retinoids in the photoreceptor cells, for example rhodopsin, the cone opsins, and ABCR, are also relevant to pathogenesis associated with defects in the visual cycle. The reader is referred to the corresponding chapters in this book for discussion of these disease genes.

RPE65

RPE65 (RPE-specific protein, 65 kD) (MIM 180069) encodes an abundant, RPE-specific, evolutionarily conserved protein that is peripherally associated with the RPE smooth endoplasmic reticulum [8, 9]. Studies of *Rpe65* knockout mice established that RPE65 is required for the conversion of vitamin A to 11-*cis* retinal [10]. In the absence of Rpe65, all-*trans* retinyl esters accumulate in droplets

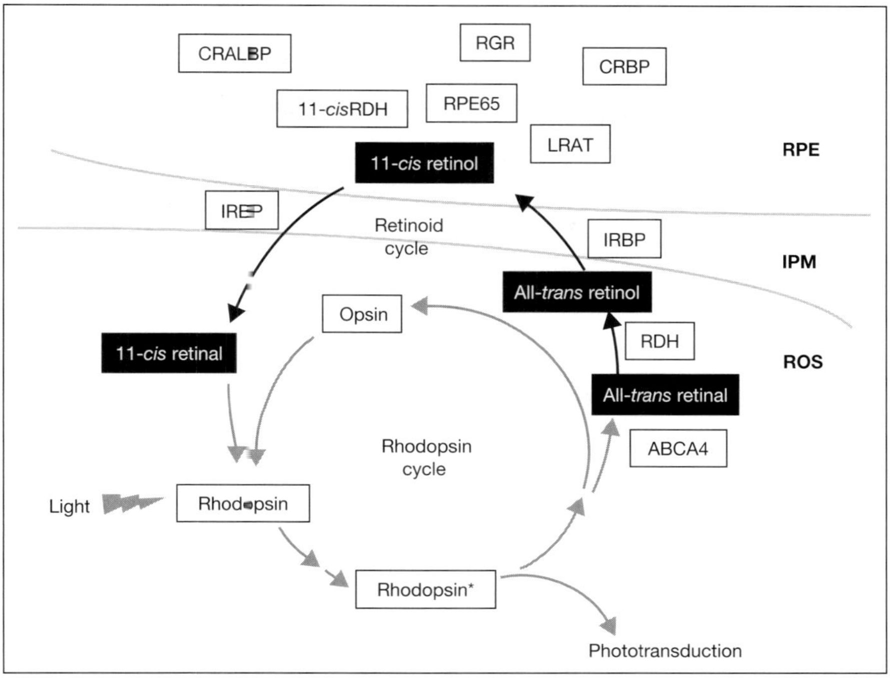

Fig. 1. The visual cycle of retinoid processing and trafficking in the outer retina. Abbreviations for protein names (in boxes) are as in the text. RPE, retinal pigment epithelium; IPM, interphotoreceptor matrix; ROS, rod outer segment; Rhodopsin*, photo activated rhodopsin. Adapted from [6].

within RPE cells, suggesting that retinoid processing is blocked following esterification of vitamin A to membrane lipids. Addition of recombinant RPE65 to RPE microsomal membranes prepared from *Rpe65*−/− mice restores the retinoid isomerase activity to the preparations [11]. *Rpe65*−/− mice do not generate the rhodopsin photopigment and have severely depressed electroretinogram (ERG) responses [10]. Remaining visual capacity is attributed to residual rod function having reset sensitivity [12] and proposed to be sustained by small amounts of 11-*cis* retinal generated by photoisomerization [13]. The mice also exhibit decreased accumulation of lipofuscin, a lipid-retinoid storage product derived from ingested photoreceptor outer segments [14], as well as increased levels of phosphorylated opsin that may represent a potential link to downstream pathogenic events [15].

Functional deficits and ultrastructural abnormalities similar to those present in *Rpe65*−/− mice are also present in a strain of Swedish briard dogs carrying a functional null allele of *Rpe65* that are afflicted with congenital night-blindness

Table 1. Disease genes involved in vitamin A metabolism of the RPE

	Chromosomal localization	Exon number	Protein size	Protein function/activity	Primary phenotype
RPE65	1p31	14	533 aa	Metabolism of all-*trans* retinyl esters to 11-*cis* retinol	(ar) childhood-onset severe retinal dystrophy (Leber amaurosis)
RLBP1	15q26	8 (7)	317 aa	Binds 11-*cis* retinol/retinal, stimulates RDH5 activity	(ar) retinitis punctata albescens
RDH5	12q13–q14	5 (4)	318 aa	Converts 11-*cis* retinol to 11-*cis* retinal	(ar) fundus albipunctatus
RGR	10q23	7	291 aa	Light-dependent isomerization of all-*trans* retinal to 11-*cis* retinal	(ar) retinitis pigmentosa (ad) choroidal sclerosis
LRAT	4q31	3 (2)	230 aa	Esterification of all-*trans* retinol to membrane phospholipid	(ar) childhood-onset severe retinal dystrophy (Leber amaurosis)

Numbers in parentheses denote coding exons. aa, amino acids; ar, autosomale recessive; ad, autosomale dominant.

associated with a progressive retinal dystrophy [16, 17]. Affected dogs derived from this breed were recently used in the first successful gene replacement therapy experiments in a large animal model of retinal degeneration [18].

Mutations in *RPE65* were initially identified in patients with autosomal recessive severe retinal dystrophy [19] and Leber congenital amaurosis type II (LCA II) [20, 21]. Since then, about 60 different disease-associated *RPE65* mutations have been described, including missense and nonsense point mutations, splice-site defects, and rearrangements affecting a few nuclestides in all 14 exons of the gene [22 and references therein]. Missense and presumed null mutations occur at approximately the same frequency. All missense mutations affect amino acid residues that are conserved across species, but are not predicted to disrupt protein folding or posttranslational processing. It is currently estimated that *RPE65* mutations are responsible for approximately 2% of all autosomal recessive retinal dystrophy alleles, and approximately 11% of all autosomal recessive severe, childhood-onset retinal dystrophy alleles [22].

The first retinal dystrophy case resulting from uniparental disomy was found to be due to complete paternal isodisomy for chromosome 1 and reduction to homoallelism for a presumed loss-of-function allele of *RPE65* [23].

Evaluation of available clinical descriptions suggests that the disease phenotype associated with *RPE65* mutations is relatively uniform overall, and largely independent of mutation type. This phenotype is characterized by poor but useful visual function in early life (measurable cone ERGs) that declines dramatically throughout the school age years, although a number of patients retain useful, albeit considerably compromised residual islands of vision into the third decade of life. It has been proposed that LCA patients with *RPE65* mutations can be distinguished, on clinical grounds, from patients who have mutations in the photoreceptor-specific guanylate cyclase gene, *GuCYiD* [24].

RLBP1

The cellular retinaldehyde-binding protein (CRALBP) encoded by *RLBP1* (MIM 180090) is an abundant retinoid-binding protein present in RPE and Müller cells where it interacts with 11-*cis* retinol and 11-*cis* retinal [25]. Studies of retinoid processing in wild-type and *Rlbp1* knockout mice show that the apoprotein stimulates the enzymatic conversion of all-*trans* retinol to 11-*cis* retinol [26]. Compared to wild-type mice, *Rlbp1*$-/-$ mice exhibit $>$10 fold delays in rates of rhodopsin regeneration, 11-*cis* retinal production, and dark adaptation. However, *Rlbp1*$-/-$ mice exhibit normal photosensitivity and show no evidence of photoreceptor degeneration at ages up to 1 year old, suggesting that another protein substitutes for CRALBP function, or that the low levels of 11-*cis* retinal are tolerated by the mouse retina.

Mutations in *RLBP1* are relatively rare and associated primarily with retinitis punctata albescens [27]. Retinitis punctata albescens is a flecked retinal dystrophy, characterized by early-onset night blindness, elevated dark adaptation thresholds, and uniform white dots across the fundus. First symptoms are most often diagnosed in young adults, and progress to a generalized atrophy of the macula and retina that results in legal blindness in early or mid adulthood. An *RLBP1* missense mutation (R234W) was identified in patients with Bothnia dystrophy, a type of retinitis punctata albescens resulting from a founder effect in a small region of Sweden [28]. *RLBP1* mutations were later identified in patients with Newfoundland rod-cone dystrophy, a severe form of retinitis punctata albescens present in a small region of Canada [29].

The first reported mutation in *RLBP1* was a missense mutation R150Q found in homozygous form in all affected individuals of a single family [30]. Studies of recombinant CRALBP protein expressed in *Escherichia coli* showed that the R150Q mutant had reduced solubility and did not bind 11-*cis* retinal in in vitro assays. Although the phenotype was in this family not described in detail, the disease associated with this mutation was reported as retinitis pigmentosa. The ophthalmological findings, however, describe the presence of macular degeneration and small white dots scattered across the fundus, and are potentially consistent with an alternative diagnosis of retinitis punctata albescens. The same R150Q missense mutation in homozygous form was also found to be responsible for late-onset, slowly progressing disease in a consanguineous kindred from Saudi Arabia [31]. Young patients in this family (3–20 years old) were initially diagnosed with fundus albipunctatus, as their fundi were covered with yellow-white flecks, and they exhibited delayed dark adaptation as well as reduced ERG responses, but had no other signs of retinal pathology. However, older family members in the kindred (40–50 years old) with the same mutation exhibited symptoms fully consistent with a diagnosis of retinitis punctata albescens, including progressive loss of cone function and generalized atrophy of the retina.

RDH5

RDH5 encodes the enzyme 11-*cis* retinol dehydrogenase (11cisRDH) (MIM 601617) that catalyzes the NADP$^+$-dependent oxidation of 11-*cis* retinol to the aldehyde, 11-*cis* retinal [32]. 11cisRDH is a homodimeric integral membrane protein present in the RPE smooth endoplasmic reticulum and plasma membrane. Within the eye, *RDH5* expression is restricted to the RPE, whereas it is expressed at lower levels in a number of tissues outside the eye [33]. 11cisRDH is a member of the family of short chain dehydrogenases/reductases that act on various hydrophobic substrates. A number of lines of evidence suggest that 11cisRDH forms a functional complex in vivo with other proteins involved in RPE retinoid processing, including RPE65 [32], RGR [34] and CRALBP [35].

Rdh5 knockout mice show delayed dark adaptation, but only at high levels of bleach that are much greater than those needed to detect abnormalities in patients with *RDH5* mutations [36]. The mice also accumulate 11-*cis* retinyl/13-*cis* retinyl esters in the RPE, suggesting that 11cisRDH activity is important for eliminating 13-*cis* retinoid catabolic intermediates. However, *Rdh5*$-/-$ mice exhibit normal fundus appearance and have no profound functional deficits, suggesting that there may be an alternative oxidative mechanism in mouse RPE that compensates for loss of 11cisRDH activity.

RDH5 disease-associated mutations were first identified in individuals diagnosed with fundus albipunctatus [37]. Fundus albipunctatus is a rare form of stationary night blindness characterized by delayed rod and cone pigment regeneration after exposure to a strong bleaching light, and a fundus covered by a myriad of tiny white flecks extending from the macula to the midperiphery. Subsequent studies showed that RDH5 mutations are a major cause of fundus albipunctatus [38, 39]. To date, a total of 18 different disease-associated mutations of *RDH5* have been reported, including 14 missense substitutions, all in patients with fundus albipunctatus [40 and references therein]. Biochemical studies of 11 of the missense mutations showed evidence of decreased protein stability, subcellular mislocalization, and (for all but A294P) decreased enzymatic activity [40]. The A294P protein is catalytically active, but subject to inactivation by dimerization with non-functional alleles. The wild-type enzyme is not subject to similar dominant negative down-regulation, consistent with observed normal phenotypes of carriers. The relatively mild phenotype and nonprogressive nature of the disease associated with mutations in *RDH5* suggests that, as in mouse, compensatory oxidative mechanisms may exist in human RPE.

RGR

RGR encodes a light-sensitive rhodopsin homolog, RPE-retinal G-protein-coupled receptor (MIM 600342), present in RPE and Müller cells [41]. The RGR protein preferentially binds all-*trans* retinal through a covalent Schiff base linkage with a lysine residue that is the counterpart of the retinal attachment site in rhodopsin. In vitro studies show that illumination of RGR results in the conversion of bound all-*trans* retinal to 11-*cis* retinal, the inverse of the light-induced rhodopsin isomerization reaction [42]. Studies of *Rgr* knockout mice show that RGR is necessary for maintaining normal steady-state levels of 11-*cis* retinal and rhodopsin photopigment in ambient light [43]. *Rgr−/−* mice have normal dark adapted levels of 11-*cis* retinal and rhodopsin photopigment, indicating that retinoid isomerase activity as well as mechanisms of dark adaptation remain functional. In addition, there are no signs of retinal degeneration in animals up to 9 months of age. However, when *Rgr−/−* mice are exposed to light, there is a drop in the levels of 11-*cis* retinal and rhodopsin, and a precursor pool of all-*trans* retinyl esters accumulates. RGR thus appears to play a predominant role in rhodopsin regeneration under photopic conditions.

Bovine RGR copurifies with a functional form of 11cisRDH that shows specificity for 11-*cis* retinal as substrate and NADH as cofactor, and converts the product of the RGR light-dependent reaction to 11-*cis* retinol [34]. The significance of the synthesis of 11-*cis* retinol in the RPE is not known, as it cannot regenerate rhodopsin, but can regenerate the cone pigments and thereby potentially impact the cone visual processing. Alternatively, synthesis of 11-*cis* retinol may play an important role in the storage of this isomeric form as 11-*cis* retinal esters, or affect the functional interactions of CRALBP.

Mutations in *RGR* are a rare cause of autosomal recessive retinitis pigmentosa, initially detected by screening a large group of patients (842 individuals) with various forms of inherited retinal disease [44]. A homozygous S66R missense mutation was found in 1 patient diagnosed with autosomal recessive retinitis pigmentosa, as well as in her 4 affected siblings, all of whom exhibited symptoms typical of adult-onset disease. In addition, a heterozygous 1-bp insertion in codon 275 was found in a second patient who had 2 affected heterozygous siblings, and a deceased affected father. The 1-bp insertion results in a frame shift that converts final 16 residues of RGR into 77 residues of novel sequence. Much remains to be understood about recessive vs. dominant mode of transmission of the trait in the two families and, in general, about the role of RGR in the visual cycle.

LRAT

LRAT (lecithin retinol acyltransferase) (MIM 604863) encodes an integral membrane protein expressed at high levels in the RPE, liver, and intestine, tissues that play critical roles in the uptake, storage and metabolism of vitamin A [45]. LRAT catalyzes the esterification of vitamin A to membrane phospholipids to produce all-*trans* retinyl esters in a reaction analogous to that performed by lecithin cholesterol acyltransferase (LCAT), a component of serum HDL particles important for lipoprotein metabolism [46]. However, analysis of the coding sequence showed that LRAT is unique and structurally unrelated to the LCAT family of proteins. In the RPE, all-*trans* retinyl esters are proposed to serve as substrates for the all-*trans* to 11-*cis* isomerization reaction [7], as well as constitute a sink for vitamin A uptake and accumulation [47].

Mutations in *LRAT* were identified by screening a group of retinal dystrophy patients (267 individuals) [48]. A homozygous S175R missense mutation that segregated with severe and early-onset retinal degeneration was present in two consanguineous, but apparently unrelated families. The S175R substitution was shown to inactivate the retinol acyltransferase activity of the recombinant protein in transfected COS-7 cells, suggesting that retinal degeneration results from homozygosity for a *LRAT* non-functional allele. The phenotype associated with this *LRAT* mutation is similar to that associated with mutations in *RPE65*. In the case of *RPE65* mutations, it is not known whether the disease results directly from loss of 11-*cis* retinal or from accumulation of all-*trans* ester intermediates, or both. In contrast, LRAT acts at or near the first step in the pathway, suggesting that in the case of *LRAT* mutations, loss of the chromophore alone may be sufficient to initiate disease pathogenesis.

Future Prospects for Research and Therapy

The discovery that defects in retinal vitamin A metabolism are responsible for inherited retinal disease established for the first time that 11-*cis* retinal is required not only for the proper function but also for the health and survival of the photoreceptor cells, and created a new focus for research into the causes and treatments of this group of diseases. Current efforts are focused on elucidating the detailed mechanism by which vitamin A is converted to 11-*cis* retinal, defining the full spectra of genetic defects and phenotypes associated with mutations in this pathway, and determining the best strategies for therapeutic intervention.

At this time, there are a number of critical gaps in our understanding of vitamin A processing at the molecular level. Questions remain concerning the identity of the retinoid isomerase and the precise role of RPE65 in isomerase activity. Physiological roles for a number of known RPE retinoid-binding proteins, including RGR and peropsin [49], have not been established. Little is known about the mechanism(s) by which retinoids traffic through the subretinal space, and enter and exit the RPE and photoreceptors. In addition, a number of proteins responsible for various known retinoid-processing activities, such as 11-*cis* retinyl ester hydrolase [50], have not been identified.

The potential impact of defects in vitamin A metabolism on phototransduction suggests that mutations at any point in the visual cycle are likely to result in inherited retinal disease. Thus, as additional key proteins and enzymes involved in vitamin A metabolism are identified, each will represent an important new candidate disease gene for consideration. It will also be important to consider the potential involvement of other known genes in this pathway, including interstitial retinol-binding protein [51], photoreceptor all-*trans* retinol dehydrogenases [52], and peropsin [49]. For some of these candidates, initial screening studies have not resulted in the identification of disease-causing mutations, suggesting that defects in these genes may be rare or restricted to certain unique phenotypes [e.g. 53].

With respect to phenotype, it is interesting to note that macular involvement is a characteristic of the disease caused by mutations in *RLBP1*, that cone dystrophy is present in many elderly patients with mutations in *RDH5*, and that macular holes are present in some patients with *RPE65* mutations by the third decade of life [54]. Although no clear links have yet been established, the notion that certain genes primarily associated with monogenic disease may also play a role in multifactorial diseases of later onset, including age-related macular degeneration (ARMD), remains an attractive hypothesis that is the focus of considerable current research effort. Differences in age of onset could result from differences in mutation type or number, variable penetrance, and/or genetic background. Alternatively, sequence variants present in a number of genes that act collectively to decrease throughput of a pathway could contribute to increased susceptibility. In light of the evidence pointing toward the early involvement of the RPE in ARMD, it will be of interest to consider the potential contribution of genetic variability in vitamin A metabolism to ARMD susceptibility, as well as the appropriateness of treatment strategies focused on vitamin A metabolism.

Defects in vitamin A metabolism appear certain to play a pivotal role in establishing the first successful paradigms for targeted therapeutic intervention in retinal dystrophies. The RPE is relatively accessible to physical and functional manipulation, and replacement strategies are likely to be appropriate in the case of autosomal recessive diseases. Initial studies using *Rpe65*−/− mice tested retinoid administration as an approach for the treatment of defects in this pathway. A single oral dose of 9-*cis* retinal resulted in the formation of the photopigment isorhodopsin and restored ERG responses in the knockout mice [13]. Early administration of repeated doses of 9-*cis* retinal reduced retinyl ester accumulation in the RPE and supported rod retinal function for more than 6 months [15]. However, many questions remain regarding the potential usefulness of this approach for patients, and future studies are needed to address dosage and toxicity issues, species differences, delivery modalities, and effects on disease pathogenesis.

A second set of studies tested gene replacement therapy in the Swedish briard dog, a naturally occurring model of the human disease caused by mutations in *RPE65*, and achieved results that electrified the field. A single subretinal injection of an AAV construct containing canine *Rpe65* restored dark-adapted ERG responses, photopigment formation and psychophysical outcomes [18]. Additional studies showed that the magnitude of the effect of a single treatment persisted for up to 13 months, assessed by ERG and behavioral testing [55]. Continued work is aimed toward establishing a platform from which to launch phase I/II clinical trials, addressing issues related to vectors, tissue specificity, toxic effects and therapeutic results.

Future tests of more general approaches are certain to follow, including RPE transplantation and treatment with survival factors. The successful application of these approaches to defects in vitamin A metabolism would represent an important advance for strategies developed over many years and with great effort. Continued progress toward the development of additional treatments will require a deeper understanding of the mechanisms of pathogenesis and degeneration, as well as specialized cell biology. In addition, necessary clinical trials will require the large-scale ascertainment of patients and development of better outcomes measures. It will also be important to define the full range of phenotypes attributable to mutations in this pathway, as they may account for a broader clinical spectrum of diagnoses (and numbers of patients) than currently appreciated. It is hoped that the results of such research efforts will mark the beginning of an era of optimism, treatment and cure for this group of devastating diseases in both young and aging patients.

Acknowledgements

The authors thank the following agencies for support of their work cited in this chapter: Deutsche Forschungsgemeinschaft, The National Institutes of Health (National Eye Institute), USA, The Foundation Fighting Blindness, USA, The British Retinitis Pigmentosa Society, The American Health Assistance Foundation, and The Midwest EyeBank and Transplantation Center.

References

1 Wald G: Molecular basis of visual excitation. Science 1968;162:230–239.

2 Fulton BS, Rando RR: Biosynthesis of 11-*cis*-retinoids and retinyl esters by bovine pigment epithelium membranes. Biochemistry 1987;26:7938–7945.

3 Sun H, Molday RS, Nathans J: Retinal stimulates ATP hydrolysis by purified and reconstituted ABCR, the photoreceptor-specific ATP-binding cassette transporter responsible for Stargardt disease. J Biol Chem 1999;274:8269–8281.

4 Weng J, Mata NL, Azarian SM, Tzekov RT, Birch DG, Travis GH: Insights into the function of Rim protein in photoreceptors and etiology of Stargardt's disease from the phenotype in abcr knockout mice. Cell 1999;98:13–23.

5 Saari JC: Biochemistry of visual pigment regeneration. The Friedenwald Lecture. Invest Ophthalmol Vis Sci 2000;41:337–348.

6 McBee JK, Palczewski K, Baehr W, Pepperberg DR: Confronting complexity: The interlink of phototransduction and retinoid metabolism in the vertebrate retina. Prog Retin Eye Res 2001;20:469–529.

7 Deigner PS, Law WC, Canada FJ, Rando RR: Membranes as the energy source in the endergonic transformation of vitamin A to 11-*cis*-retinol. Science 1989;244:968–971.

8 Hamel CP, Tsilou E, Pfeffer BA, Hooks JJ, Detrick B, Redmond TM: Molecular cloning and expression of RPE65, a novel retinal pigment epithelium-specific microsomal protein that is post-transcriptionally regulated in vitro. J Biol Chem 1993;268:15751–15757.

9 Nicoletti A, Wong DJ, Kawase K, Gibson LH, Yang-Feng TL, Richards JE, Thompson DA: Molecular characterization of the human gene encoding an abundant 61-kDa protein specific to the retinal pigment epithelium. Hum Mol Genet 1995;4:641–649.

10 Redmond TM, Yu S, Lee E, Bok D, Hamasaki D, Chen N, Goletz P, Ma JX, Crouch RK, Pfeifer K: *Rpe65* is necessary for production of 11-*cis*-vitamin A in the retinal visual cycle. Nat Genet 1998;20:344–351.

11 Crouch RK, Moiseyev G, Gloletz P, Bealle G, Redmond TM, Ma JX: RPE65 is essential but not sufficient for production of 11-*cis* retinal. Invest Ophthalmol Vis Sci 2001;42:S655, ARVO abstr 3525.

12 Seeliger MW, Grimm C, Stahlberg F, Friedburg C, Jaissle G, Zrenner E, Guo H, Reme CE, Humphries P, Hofmann F, Biel M, Fariss RN, Redmond TM, Wenzel A: New views on RPE65 deficiency: The rod system is the source of vision in a mouse model of Leber congenital amaurosis. Nat Genet 2001;29:70–74.

13 Van Hooser JP, Liang Y, Maeda T, Kuksa V, Jang GF, He YG, Rieke F, Fong HK, Detwiler PB, Palczewski K: Recovery of visual functions in a mouse model of Leber congenital amaurosis. J Biol Chem 2002;277:19173–19182.

14 Katz ML, Redmond TM: Effect of Rpe65 knockout on accumulation of lipofuscin fluorophores in the retinal pigment epithelium. Invest Ophthalmol Vis Sci 2001;42:3023–3030.

15 Van Hooser JP, Aleman TS, He YG, Cideciyan AV, Kuksa V, Pittler SJ, Stone EM, Jacobson SG, Palczewski K: Rapid restoration of visual pigment and function with oral retinoid in a mouse model of childhood blindness. Proc Natl Acad Sci USA 2000;97:8623–8628.

16 Wrigstad A, Narfstrom K, Nilsson SE: Slowly progressive changes of the retina and retinal pigment epithelium in Briard dogs with hereditary retinal dystrophy: A morphological study. Doc Ophthalmol 1994;87:337–354.

17 Veske A, Nilsson SD, Narfström K, Gal A: Retinal dystrophy of Swedish Briard/Briard-beagle dogs is due to a 4-bp deletion in *RPE65*. Genomics 1999;57:57–61.

18 Acland GM, Aguirre GD, Ray J, Zhang Q, Aleman TS, Cideciyan AV, Pearce-Kelling SE, Anand V, Zeng Y, Maguire AM, Jacobson SG, Hauswirth WW, Bennett J: Gene therapy restores vision in a canine model of childhood blindness. Nat Genet 2001;28:92–95.

19 Gu SM, Thompson DA, Srikumari CRS, Lorenz B, Finckh U, Nicoletti A, Murthy KR, Rathmann M, Kumaramanickavel G, Denton MJ, Gal A: Mutations in *RPE65* cause autosomal recessive childhood-onset severe retinal dystrophy. Nat Genet 1997;17:194–197.

20 Marlhens F, Bareil C, Griffoin JM, Zrenner E, Amalric P, Eliaou C, Liu SY, Harris E, Redmond TM, Arnaud B, Claustres M, Hamel CP: Mutations in *RPE65* cause Leber's congenital amaurois. Nat Genet 1997;17:139–141.

21 Morimura H, Fishman GA, Grover SA, Fulton AB, Berson EL, Dryja TP: Mutations in the *RPE65* gene in patients with autosomal recessive retinitis pigmentosa or Leber congenital amaurosis. Proc Natl Acad Sci USA 1998;95:3088–3093.

22 Thompson DA, Gyurus P, Fleischer L, Bingham EL, McHenry CL, Apfelstedt-Sylla E, Zrenner E, Lorenz B, Richards JE, Jacobson SG, Sieving PA, Gal A: Genetics and phenotypes of RPE65 mutations in inherited retinal degeneration. Invest Ophthalmol Vis Sci 2000;41:4293–4299.

23 Thompson DA, McHenry CL, Li Y, Richards JE, Othman MI, Schwinger E, Vollrath D, Jacobson SG, Gal A: Retinal dystrophy due to paternal isodisomy for chromosome 1 or chromosome 2, with homoallelism for mutations in RPE65 or MERTK, respectively. Am J Hum Genet 2002;70:224–229.

24 Perrault I, Rozet JM, Ghazi I, Leowski C, Bonnemaison M, Geber S, Ducroq D, Cabot A, Souied E, Dufier JL, Munnich A, Kaplan J: Different functional outcome of *RetGC1* and *RPE65* gene mutations in Leber congenital amaurosis. Am J Hum Genet 1999;64:1225–1228.

25 Intres R, Goldflam S, Cook JR, Crabb JW: Molecular cloning and structural analysis of the human gene encoding cellular retinaldehyde-binding protein. J Biol Chem 1994;269:25411–25418.

26 Saari JC, Nawrot M, Kennedy BN, Garwin GG, Hurley JB, Huang J, Possin DE, Crabb JW: Visual cycle impairment in cellular retinaldehyde-binding protein (CRALBP) knockout mice results in delayed dark adaptation. Neuron 2001;29:739–748.

27 Morimura H, Berson EL, Dryja TP: Recessive mutations in the RLBP1 gene encoding cellular retinaldehyde-binding protein in a form of retinitis punctata albescens. Invest Ophthalmol Vis Sci 1999;40:1000–1004.

28 Burstedt MS, Sandgren O, Holmgren G, Forsman-Semb K: Bothnia dystrophy caused by mutations in the cellular retinaldehyde-binding protein gene (RLBP1) on chromosome 15q26. Invest Ophthalmol Vis Sci 1999;40:995–1000.

29 Eichers ER, Green JS, Stockton DW, Jackman CS, Whelan J, McNamara JA, Johnson GJ, Lupski JR, Katsanis N: Newfoundland rod-cone dystrophy, an early-onset retinal dystrophy, is caused by splice-junction mutations in RLBP1. Am J Hum Genet 2002;70:955–964.

30 Maw MA, Kennedy B, Knight A, Bridges R, Roth KE, Mani EJ, Mukkadan JK, Nancarrow D, Crabb JW, Denton MJ: Mutation of the gene encoding cellular retinaldehyde-binding protein in autosomal recessive retinitis pigmentosa. Nat Genet 1997;17:198–200.

31 Katsanis N, Shroyer NF, Lewis RA, Cavender JC, Al-Rajhi AA, Jabak M, Lupski JR: Fundus albipunctatus and retinitis punctata albescens in a pedigree with an R150Q mutation in RLBP1. Clin Genet 2001;59:424–429.
32 Simon A, Hellman U, Wernstedt C, Eriksson U: The retinal pigment epithelial-specific 11-*cis* retinol dehydrogenase belongs to the family of short chain alcohol dehydrogenases. J Biol Chem 1995;270:1107–1112.
33 Wang J, Chai X, Eriksson U, Napoli JL: Activity of human 11-*cis*-retinol dehydrogenase (Rdh5) with steroids and retinoids and expression of its mRNA in extraocular human tissue. Biochem J 1999;338:23–27.
34 Chen P, Lee TD, Fong HK: Interaction of 11-*cis*-retinol dehydrogenase with the chromophore of retinal G-protein-coupled receptor opsin. J Biol Chem 2001;276:21098–21104.
35 Bhattacharya SK, Wu Z, Miyagi M, West KA, Jin Z, Nawrot M, Saari JC, Crabb JW: Interactions of CRALBP with other visual cycle proteins. Invest Ophthalmol Vis Sci 2002, ARVO abstr 4567.
36 Driessen CAGG, Winkens HJ, Hoffmann K, Kuhlmann LD, Janssen BPM, Van Vugt AHM, Van Hooser JP, Wieringa BE, Deutman AF, Palczewski K, Ruether K, Janssen JJM: Disruption of the 11-*cis*-retinol dehydrogenase gene leads to accumulation of *cis*-retinols and *cis*-retinyl esters. Mol Cell Biol 2000;20:4275–4287.
37 Yamamoto H, Simon A, Eriksson U, Harris E, Berson EL, Dryja TP: Mutations in the gene encoding 11-*cis* retinol dehydrogenase cause delayed dark adaptation and fundus albipunctatus. Nat Genet 1999;22:188–191.
38 Gonzalez-Fernandez F, Kurz D, Bao Y, Newman S, Conway BP, Young JE, Han DP, Khani SC: 11-*cis* Retinal dehydrogenase mutations as a major cause of the congenital night-blindness disorder known as fundus albipunctatus. Mol Vis 1999;5:41.
39 Hirose E, Inoue Y, Morimura H, Okamoto N, Fukuda M, Yamamoto S, Fujikado T, Tano Y: Mutations in the 11-*cis* retinol dehydrogenase gene in Japanese patients with fundus albipunctatus. Invest Ophthalmol Vis Sci 2000;41:3933–3935.
40 Lidén M, Romert A, Tryggvason K, Persson B, Eriksson U: Biochemical defects in 11-*cis* retinol dehydrogenase mutants associated with fundus albipunctatus. J Biol Chem 2001;256: 49251–49257.
41 Shen D, Jiang M, Hao W, Tao L, Salazar M, Fong HKW: A human opsin-related gene that encodes a retinaldehyde-binding protein. Biochemistry 1994;33:13117–13125.
42 Hao W, Fong HK: The endogenous chromophore of retinal G-protein-coupled receptor opsin from the pigment epithelium. J Biol Chem 1999;274:6085–6090.
43 Chen P, Hao W, Rife L, Wang XP, Shen D, Chen J, Ogden T, Van Boemel GB, Wu L, Yang M, Fong HKW: A photic visual cycle of rhodopsin regeneration is dependent on Rgr. Nat Genet 2001;28: 256–260.
44 Morimura H, Saindelle-Ribeaudeau F, Berson EL, Dryja TP: Mutations in RGR, encoding a light-sensitive opsin homologue, in patients with retinitis pigmentosa. Nat Genet 1999;23:393–394.
45 Ruiz A, Winston A, Lim YH, Gilbert BA, Rando RR, Bok D: Molecular and biochemical characterization of lecithin retinol acyltransferase. J Biol Chem 1999;274:3834–3841.
46 McLean J, Fielding C, Drayna D, Dieplinger H, Baer B, Kohr W, Henzel W, Lawn R: Cloning and expression of human lecithin-cholesterol acyltransferase cDNA. Proc Natl Acad Sci USA 1986;83:2335–2339.
47 McBee JK, Kuksa V, Alvarez R, de Lera AR, Prezhdo O, Haeseleer F, Sokal I, Palczewski K: Isomerization of all-*trans*-retinol to *cis*-retinols in bovine retinal pigment epithelial cells: Dependence on the specificity of retinoid-binding proteins. Biochemistry 2000;39:11370–11380.
48 Thompson DA, Li Y, McHenry CL, Carlson TJ, Ding X, Sieving PA, Apfelstedt-Sylla E, Gal A: Mutations in the gene encoding lecithin retinol acyltransferase are associated with early-onset severe retinal dystrophy. Nat Genet 2001;28:123–124.
49 Sun H, Gilbert DJ, Copeland NG, Jenkins NA, Nathans J: Peropsin, a novel visual pigment-like protein located in the apical microvilli of the retinal pigment epithelium. Proc Natl Acad Sci USA 1997;94:9893–9898.
50 Mata NL, Villazana ET, Tsin AT: Colocalization of 11-*cis* retinyl esters and retinyl ester hydrolase activity in retinal pigment epithelium plasma membrane. Invest Ophthalmol Vis Sci 1998;39: 1312–1319.

51 Liou GI, Ma DP, Yang YW, Geng L, Zhu C, Baehr W: Human interstitial retinoid-binding protein: Gene structure and primary sequence. J Biol Chem 1989;264:8200–8206.
52 Rattner A, Smallwood PM, Nathans J: Identification and characterization of all-*trans*-retinol dehydrogenase from photoreceptor outer segments, the visual cycle enzyme that reduces all-*trans*-retinal to all-*trans*-retinol. J Biol Chem 2000;275:11034–11043.
53 Valverde D, Vazquez-Gundin F, del Rio E, Calaf M, Fernandez JL, Baiget M: Analysis of the IRBP gene as a cause of RP in 45 ARRP Spanish families. Ophthalmic Genet 1998;19:197–202.
54 Felius J, Thompson DA, Khan NW, Bingham EL, Jamison JA, Kemp JA, Sieving PA: Clinical course and visual function in a family with mutations in the RPE65 gene. Arch Ophthalmol 2002;120:55–61.
55 Acland GM, Aguirre GD, Aleman TS, Cideciyan AV, Maguire AM, Jacobson SG, Hauswirth WW, Bennett J: Continuing evaluation of gene therapy in the Rpe65 mutant dog. Invest Ophthalmol Vis Sci 2002, ARVO abstr 4593.

Andreas Gal, MD, PhD
Institut für Humangenetik, Universitätsklinikum Hamburg-Eppendorf
Butenfeld 42, D–22529 Hamburg (Germany)
Tel. +49 40 42803 2120, Fax +49 40 42803 5138, E-Mail gal@uke.uni-hamburg.de

Wissinger B, Kohl S, Langenbeck U (eds): Genetics in Ophthalmology.
Dev Ophthalmol. Basel, Karger, 2003, vol 37, pp 155–169

Genetics of Macular Dystrophies and Implications for Age-Related Macular Degeneration

Caroline C.W. Klaver[a,c], *Rando Allikmets*[a,b]

Departments of [a]Ophthalmology and [b]Pathology, Columbia University, New York, N.Y., USA and [c]Department of Ophthalmology, Erasmus University, Rotterdam, The Netherlands

Abstract

Determining the genetic component of the age-related macular degeneration (AMD) complex trait has been the primary goal of ophthalmic genetics research for almost a decade. During this time, genes of several Mendelian traits affecting the macula have been identified. In this review, we will discuss the consequences of molecular defects in the VMD2, EFEMP1, TIMP3, ELOVL4 and ABCA4 genes, and their association with macular disease. We will also analyze our current knowledge on the implications of genetic variations in these genes for AMD by summarizing data from all studies which have investigated the possible role of these candidate genes in the etiology of AMD. Finally, we will elaborate on methods for genetic dissection of complex traits and discuss the appropriate applications of these methods for identifying genetic determinants of AMD.

Introduction

Age-related macular degeneration (AMD) is the most common cause of acquired visual impairment in people over 60 years of age, and is estimated to affect millions of individuals in the Western world. Prevalence increases with age; among persons aged 75 years and older, mild, or early forms occur in nearly 30% and advanced forms in about 7% of the population [1]. The current understanding is that AMD represents a multifactorial disorder involving both environmental and genetic factors. Environmental risk factors which have been associated with the disorder include smoking, hypertension, diet and exposure to sunlight, although findings are not unequivocal among studies [1].

Familial aggregation, segregation and twin studies have clearly established that genetic predisposition plays a major role in the etiology of AMD [2, 3]. The lifetime relative risk of advanced AMD for first-degree relatives is estimated to be 4.2, indicating that they are 4 times more likely to develop the disorder than non-related subjects [2]. The population-attributable risk is calculated to be 23%, suggesting that the proportion of AMD directly related to genes amounts to approximately one quarter of all cases. It is assumed that multiple genes are involved, although the approximate number of those has not been reliably estimated. Furthermore, as in other complex traits, the extent of genetic heterogeneity cannot be estimated so that influences of many genes and their interactions are difficult to ascertain even in individual families. In summary, AMD is yet another example of a complex trait where the genetic studies are complicated by the very late onset of the disorder, decreased penetrance, and potential genetic heterogeneity.

Methods applicable to studies of diseases of Mendelian inheritance, i.e., variations of linkage and linkage disequilibrium analysis, have been applied to many complex disorders including AMD. As with the majority of complex traits these have not yielded robust results in elucidating the genetic components of the complex AMD trait. Recently, substantial progress has been made in determining the genetic basis of monogenic eye disorders. On a monthly basis, mutations are identified in new genes responsible for some form of retinal degeneration. Most if not all of these genes become candidates for potential involvement in multifactorial disorders especially if the phenotypes of the early-onset Mendelian diseases they cause resemble later onset complex traits. As expected, the results of case-control studies of candidate genes vary and are not always convincing.

This review attempts to summarize our current knowledge of candidate gene research aimed at delineating the molecular genetic basis of AMD. In addition, it tries to explain some of the frustration and to offer insight into what to expect from future studies.

Macular Dystrophies of Mendelian Inheritance

Best Disease

Clinical Presentation

Best disease, also known as vitelliform macular degeneration, is a congenital macular dystrophy with a wide phenotypic variation. Classical clinical presentation includes the 'egg yolk' appearance in the macula, but the disease may also develop into a pseudo-hypopyon, a disciform scar or a 'bull's eye' with central atrophy. A diagnostic hallmark is an abnormal electro-oculogram with

Table 1. Inherited macular dystrophies

Disease	Trans-mission	MIM	Locus	Gene	Expression	Protein	Protein function
Best	AD	153700	11q13	VMD2	RPE	Bestrophin	Chloride channel
Doyne	AD	126600	2p16	EFEMP1	RPE, choroid, retina, and lung, moderately in brain, heart, spleen, kidney	EGF-containing fibrillin-like extracellular matrix protein 1	Unknown Possible role in assembly of extracellular matrix
Sorsby	AD	136900	22q12.1–q13.2	TIMP3	RPE and choroid, many other tissues, including cartilage, muscle, skin, numerous epithelial layers, placenta, kidney, lung, heart, ovary, brain, and mammary tissue	Tissue inhibitor of metalloproteinase-3	Inhibition of metalloproteinases
Dominant Stargardt-like	AD	600110	6q14	ELOVL4	Photoreceptors, moderately in brain	Elongation of very long chain fatty acids protein	Unknown Possible role in synthesis of polyunsaturated fatty acids
Stargardt	AR	248200	1q21–p22.1	ABCA4	Photoreceptors	ATP-binding cassette transporter protein	Transporter of N-retinylidine-PE

an Arden ratio of <1.5, which indicates a much lower change in the electric potential derived from the RPE than normally when light is cast on the fundus. Visual symptoms such as blurred vision or metamorphopsia may occur as early as in the first decade, but significant visual decline generally does not happen before third and fourth decades of life. The latter is associated with subretinal neovascularization and central atrophy.

Genetic Defects and Functional Implications

Best disease is inherited as an autosomal dominant trait with full penetrance regarding the diminished Arden ratio. However, obligate carriers with normal fundi are relatively common. The gene, located on 11q13, was identified by positional cloning and named VMD2 in 1998 (table 1) [4]. The VMD2 gene consists of a 1,755-bp open reading frame including 11 exons and two alternative polyadenylation sites. To date, numerous sequence variants have been described in familial as well as in sporadic cases. These represent predominantly missense mutations affecting amino acids in the first half of the protein, and occur in four distinct clusters, most likely representing important functional regions [4].

The gene encodes a 585-amino-acid protein known as bestrophin, which has been localized to the basolateral plasma membrane of the RPE. The amino acid sequence is highly homologous to the RFP family in *Caenorhabditis elegans*, and most mutations affect the regions conserved in evolution from humans to *C. elegans* [4]. Most recently, bestrophin was characterized as a chloride channel, which explains the abnormal electro-oculogram in all Best disease patients [5]. In addition, experiments by Sun et al. [5] demonstrated that the dominance of mutant alleles in causing the disease is likely to result from the production of defective channels composed of both mutant and wild-type subunits.

VMD2 and Other Retinal Dystrophies

Mutations in VMD2 have been documented in a significant fraction of patients with adult vitelliform macular dystrophy, suggesting considerable allelic overlap of this late-onset dystrophy with the early-onset Best disease [6, 7]. VMD2 was screened for variants in AMD patients and matched controls by three independent groups (table 2) [6, 8]. Although none of these studies was able to document a statistically significant association of VMD2 mutations with AMD, each study did find a small, ~1–1.5%, fraction of alleles only in patient cohorts. It would be safe to say that our current knowledge of VMD2 variation in AMD excludes this gene from 'major' candidates for association with AMD. However, as discussed below, the relatively small sample size of each separate study prohibits far-reaching conclusions at this time.

Doyne Honeycomb Retinal Dystrophy

Clinical Presentation

Doyne honeycomb retinal dystrophy and malattia leventinese are currently considered allelic disorders. This condition generally begins at the end of the second decade with small macular and peripapillary drusen-like deposits, frequently distributed in a radial fashion. These lesions become larger and denser towards 30–40 years of age, while pigment epithelial atrophy and subsequent visual loss may occur during the fourth and fifth decades. The disease is not associated with specific abnormalities on psycho- and electrophysiologic tests, although non-specific aberrations have been documented.

Genetic Defects and Functional Implications

The inheritance pattern of this disease is also autosomal dominant. Most families can be traced down to origins in the Leventine Valley in southern

Table 2. Summary of Association Studies for AMD

Gene	Study	Study subjects		Findings
		cases	controls	
VMD2	Allikmets et al., 1999	259	196	No association; rare variants in 1% of cases
	Kramer et al., 2000	200		No changes found
	Lotery et al., 2000	321	192	No association; rare variants in 1.5% of cases
EFEMP1	Stone et al., 1999	494	477	No changes found
TIMP3	Felbor et al., 1997	143		One sequence change in untranslated region in 1 patient
	De la Paz et al., 1997	38 families		No evidence of linkage
	Stone et al., 2001	188		No changes found
ELOVL4	Ayyagari et al., 2001	513	292	No association; rare variants found in 5% of cases vs. 7% of controls
ABCA4	Allikmets et al., 1997	167	220	Statistically significant association with AMD – rare variants in 16% of cases vs. 1% of controls
	Stone et al., 1998	182	96	No association; non-conserved changes (including frequent polymorphisms) in 31% of cases vs. 27% of controls
	Kuroiwa et al., 1999	80	100	No association; small cohorts, incomplete screening
	De la Paz, 1999	165	56	No association; incomplete screening
	Souied et al., 2000	52 families		Two of 6 variants segregate with the disease in families; associated variants in 4% of exudative cases
	Rivera et al., 2000	200	220	Study included in the Consortium Study [Allikmets et al., 2000] No association if analyzed separately; variants in 9% of cases vs. 5% of controls
	Allikmets et al., 2000	1,218	1,258	Large independent study including data from 15 centers Significant difference in frequency of G1961E and D2177N mutations between cases (3.4%) and controls (0.95%)
	Fuse et al., 2000	25	40	No association; very small cohorts, incomplete screening
	Webster et al., 2001	182	96	Same as Stone et al., 1998; complete screening, low mutation detection rate (~25–30%), inclusion of frequent polymorphisms in the analysis

Table 2 (continued)

Gene	Study	Study subjects		Findings
		cases	controls	
	Guymer et al., 2001	544	689	No significant difference between frequency of variants G1961E and D2177N between cases (2.2%) and controls (1%)
	Allikmets 2000	1,879	1,780	Meta-analysis of all studies which had screened for G1961E and D2177N variants. Significant difference in frequencies of these variants between cases and controls (D2177N 2.0% vs. 0.5%; G1961E 1.6% vs. 0.2%).

Switzerland and northern Italy. The causal gene, mapped to chromosome 2p16 (table 1), was identified by a combination of positional cloning and candidate gene methods in 1999 as epidermal growth factor-containing fibrillin-like extracellular matrix protein 1 (EFEMP1) [9]. A single Arg345Trp mutation in exon 10 of EFEMP1 was determined to be responsible for all affected members in 37 Swiss, British and Australian families, suggesting that they are all descendents of one common ancestor [9]. The same mutation segregated with the disease in a North American family [10].

EFEMP1 encodes a 493-amino-acid extracellular matrix protein that is most abundant in eye and lung, and moderately expressed in brain, heart, spleen and kidney. In the eye, it is expressed throughout the retina, as well as in the RPE and choroid [9]. The exact protein function is still unknown, but similarities in structure and chromosomal assignment of this gene point to a homology to fibrillin and suggest a role in assembly of the extracellular matrix. The gene has six calcium-binding EGF-like domains which are hypothesized to play a key role in protein-protein interactions [11]. The Arg345Trp mutation is located in the last EGF-like domain.

EFEMP1 and Other Retinal Dystrophies

Stone et al. [9] screened the entire coding sequence of the EFEMP1 gene by SSCP in 494 patients with AMD, and found no sequence changes (table 2). Tarttelin et al. [12] screened 10 families and 17 sporadic patients with early-onset macular drusen, and found the Arg345Trp mutation in 7/10 families, and in only 1/17 of the sporadic patients. Sauer et al. [13] also screened sporadic patients with early-onset drusen and found no EFEMP1 mutations in 14 individuals.

Thus, barring phenocopies, the role of EFEMP1 mutations in dominant drusen phenotypes has to be further evaluated.

Sorsby Fundus Dystrophy

Clinical Presentation

Sorsby pseudo-inflammatory macular dystrophy is characterized by macular and extramacular chorioretinal neovascularization typically occurring in the fourth and fifth decades of life. Early features consist of small drusenoid lesions, referred to as 'colloid bodies', pigmentary clumping, and pigment epithelial atrophy, which may extend far into the periphery. Visual loss continues to progress peripherally and may deteriorate to hand motion. Color anomalies and night blindness as well as diminished rod and cone ERG photoresponses appear to be common, but are not used for definitive diagnosis.

Genetic Defects and Functional Implications

Sorsby fundus dystrophy is an autosomal dominant disease with complete penetrance. The responsible gene, found by Weber et al. [14] in 1994, encodes a tissue inhibitor of metalloproteinases (TIMP3) (table 1). To date, several mutations, including missense, nonsense and splice site mutations, have been described. These occur predominantly in the C-terminal region (exon 5) of the protein and result in conversion of the original amino acid to a cysteine, or in a deletion of most of the C-terminal region. The TIMP3 protein belongs to a family of secreted proteins that play a role in regulating extra-cellular matrix metabolism. By their ability to inhibit matrix metalloproteinases (MMPs) such as collagenases, stromelysins and gelatinases, they determine the extent of matrix degradation during normal tissue-remodeling processes. Other functions of these proteins include proMMP activation, cell growth promotion, matrix binding, inhibition of angiogenesis and the induction of apoptosis. Recent analysis of the altered proteins in SFD after incorporation of mutant TIMP3 alleles in transfected cell lines showed that mutants have normal MMP inhibitory activity, but display aberrant protein-protein interaction properties and affect cell adhesion to the extracellular matrix [15].

TIMP3 and Other Retinal Dystrophies

Assink et al. [16] and Ayyagari et al. [17] independently excluded the TIMP3 gene in two large pedigrees with autosomal dominant neovascular macular dystrophies that closely resembled Sorsby's dystrophy, suggesting that other genes can cause very similar phenotypes. Since TIMP3 was the first gene discovered of those listed here, it spurned significant interest as a potential candidate gene for AMD. First, De la Paz et al. studied 38 multiplex families with AMD and found no evidence of linkage or association between AMD and the TIMP3 locus [18]. However, as discussed below, linkage studies on a small sample of families segregating the AMD complex trait have not enough power to reach definitive conclusions. Two more studies have investigated allelic variation of TIMP3 in AMD with very similar results (table 2) [19, 20]. Both of them did not find any sequence changes in the coding region of the gene in AMD patients, suggesting that the TIMP3 locus is relatively homogeneous and not associated with AMD.

Dominant Stargardt-Like Macular Dystrophy

Clinical Presentation

Visual loss without apparent fundus lesions is a common first presentation of this disorder, usually in the first or second decade of life. The subtle early changes consist of RPE mottling and slight pallor of the optic nerve. Later, these features are followed by atrophy of the RPE in the macular area, which may or may not be accompanied by yellow flecks. Final visual acuity generally ranges from 20/40 to 20/200, the presence of yellow flecks predicting a more severe visual outcome. The 'dark choroid' on fluorescein angiography which is common in recessive Stargardt disease has not been described in the dominant form. Psycho- and electrophysical testing is mostly normal, although older patients may have diminished rod and cone amplitudes, or delayed 30-Hz flicker.

Genetic Defects and Functional Implications

The frequency of autosomal dominant Stargardt-like dystrophy is much lower than that of recessive Stargardt disease, and only a handful of families have been described. A novel photoreceptor-specific gene called ELOVL4 (elongation factor of very-long-chain fatty acids) was identified as the responsible gene by Zhang et al. [21] in 2001. To date, the only two identified disease-associated mutations in exon 6 of ELOVL4 result in the same deleterious effect on the protein: deletion of the last 51 amino acids, including a dilysine-targeting signal. The gene encodes a putative protein of 314 amino acids with approximately 35% amino acid identity to members of the ELO gene family in yeast, which encode components of the membrane-bound fatty acid elongation system. Based on the similarities, it has been hypothesized that the ELOVL4 protein is involved in synthesis of the polyunsaturated fatty acids present in the outer segments, thereby playing a crucial role in membrane composition and, therefore, photoreceptor function.

ELOVL4 and Other Retinal Dystrophies

ELOVL4 was considered a good candidate for autosomal dominant retinitis pigmentosa (RP25), since ELOVL4 maps to the same critical region. Li et al.

[22] screened eight families with 18 RP patients for mutations in ELOVL4, and did not find any associated variants. Ayyagari et al. [23] screened 778 AMD patients and 551 age-matched controls for sequence variants in ELOVL4 and detected eight amino acid-changing variants, and three SNPs in the non-coding region. Frequencies of these variants were not significantly different between cases and controls, similar to the other dominant genes described above.

Stargardt Macular Dystrophy

Clinical Presentation

Stargardt disease (STGD) is the most common macular dystrophy with an estimated frequency of 1:8,000–10,000 in the USA. The age of onset and clinical course of STGD are highly variable. One third of those affected present in the first decade of life, and they generally have a more progressive course than those with later onset. Fundus abnormalities include pigmentary changes in the macula, RPE atrophy giving a 'bull's eye' appearance, a 'beaten bronze' look of the posterior pole, and yellowish 'fishtail' flecks at the level of the RPE. The latter manifestation is also called fundus flavimaculatus. In a large fraction of STGD patients, a 'dark' or 'silent' choroid is seen on fluorescein angiography which reflects the accumulation of lipofuscin. ERG findings vary and are not diagnostic for the disease.

Genetics and Functional Implications

This macular dystrophy is the only one described here with an autosomal recessive mode of inheritance. All families segregating the disorder have been linked to chromosome 1p13–p22, confirming genetic homogeneity of the disease [24, 25]. The causal gene, ATP-binding cassette-transporter ABCA4 (ABCR), was cloned in 1997 [26]. The open reading frame of ABCA4 consists of 50 exons and encodes a 2,273-amino-acid protein, which had been previously characterized as the photoreceptor rim protein due to its localization to the rims of rod and cone outer-segment disks. The working hypothesis of ABCA4 function in vivo has the protein translocating *N-retinylidine-phosphatidylethanolamine* (N-retinylidine-PE), which forms after conversion of 11-*cis*-retinal to the all-*trans* isoform after photobleaching, across the disk membrane from the disk lumen into the cytoplasm. In the absence of a functional ABCA4 gene, N-retinylidine-PE accumulates within the outer segment disks followed by formation of N-retinylidine-N-retinylethanolamine (A2-E), the major component of lipofuscin. Consequently, abnormally high levels of lipofuscin accumulate in the RPE, triggering RPE-cell death and causing secondary photoreceptor degeneration.

The overall detection rate of disease-associated ABCA4 alleles in Stargardt disease ranges from 30 to 80%, depending on the method and the study.

The total number of identified ABCA4 mutations has grown over 400. Variants are spread across the entire gene, the majority representing missense amino acid substitutions in conserved functional domains. The heterogeneity of ABCA4 alleles is further underlined by the fact that the most frequent disease-associated variants, G1961E, G863A/delG863 and A1038V, are found in only ~10% of STGD patients. Findings of pseudo-dominant inheritance and high frequency of mutant ABCA4 alleles in the general population (~1:20) have defined ABCA4 mutations as the most frequent cause of retinal disease [27].

ABCA4 and Other Retinal Dystrophies

According to the current model, severity of retinal disease is thought to be inversely associated with residual ABCA4 protein activity. The disease phenotypes in this model range from the most severe retinitis pigmentosa (RP) affecting subjects with complete absence of ABCA4 activity due to deleterious mutations on both alleles, to AMD, which appears to occur at a higher frequency in heterozygous carriers of ABCA4 alleles.

ABCA4 involvement in AMD has been a subject of heated discussions. Various case-control association studies have come to opposing conclusions. This issue has been discussed in depth in a recent review [28]. In short, genetic studies of ABCA4 have once again revealed the complicated task of dissecting a complex trait. Studies involving exceptionally large cohorts or meta-analyses have clearly demonstrated a statistical significant association of ABCA4 variants with AMD (table 2) [28, 29]. Moreover, a recent study of a mouse model has effectively provided evidence that the absence of one ABCA4 allele can lead to retinal degeneration by demonstrating that mice heterozygous for an ABCA4 null mutation accumulated A2E in the retina and lipofuscin granules in the RPE, and exhibited delayed recovery of rod sensitivity by ERG [30].

Methods to Dissect the AMD Complex Trait

Genes associated with Mendelian diseases have been historically looked upon as good candidates for involvement in complex traits with similar phenotypic features (table 1). The same has been true for AMD – since the explosion in discovery of genes responsible for early-onset retinal dystrophies several years ago, every one of these has been extensively screened for variants in case-control association studies involving AMD patients and matched controls. This review summarizes the current status of this multi-year effort, and tries to forecast the future of similar studies.

In short, the efforts have been relatively disappointing, since none of the genes mentioned here with the exception of ABCA4 has been significantly associated with an increased susceptibility to AMD (table 2). A few common features of candidate genes are worth underlining. First, all genes, except ABCA4, are causal in diseases of autosomal dominant inheritance. Simplistically, one would think that if a dominant gene carries a mutation, the effect would result in a phenotype with a dominant Mendelian segregation. It is plausible that a fraction of variations in a dominant gene are less penetrant resulting in a delayed onset in combination with other genetic and/or environmental factors. However, a much more appealing scenario renders carriers of recessive gene variants likely to develop complications later in life. Second, in almost every gene, mutations have been found in a small, ~1%, fraction of AMD cases. Even a substantial increase in the size of study populations will not, in the vast majority of cases, result in enough power to decide if these rare variants are significantly involved in AMD. This brings us to an obvious question: Should we consider rare, infrequent variants as possibly disease-associated, or do we have enough evidence already to discard these genes as candidates for involvement in AMD?

Before we try to answer this question, we should briefly summarize our current knowledge of genetic determinants underlying complex traits, and methods applicable for studying them. Research of complex diseases has recently divided geneticists into two major 'camps' – those who believe that complex traits are caused by elevated common SNPs, and those who trust combinations of rare variants underlying these disorders. The first, 'common disease/common variant', hypothesis is supported by both theoretical analyses [31] and data from several studies (i.e., APOE in Alzheimer disease) [32]. Similarly, theoretical calculations have strongly suggested the 'rare variant' hypothesis with several examples, including the evidence gained from ABCA4 research, supporting this possibility [33]. Which of the two scenarios (or their combination) is true for AMD needs to be determined. This task is further complicated by the fact that no 'major' AMD gene (or locus) has been reliably identified. The latter brings us to the discussion about methods to dissect complex traits. These can also be divided into two major categories – linkage-based and linkage disequilibrium-based approaches. Linkage analyses, including model-free (non-parametric) sib pair approaches have served extremely well in dissecting Mendelian disorders. In complex traits they have been much less informative, especially if stringent criteria are applied [34].

Various linkage studies have been applied to confirm or reject selected candidate genes for AMD [18, 35, 36]. The outcome has been the same for all of these studies – known candidate gene loci have failed to show linkage to AMD. What these studies have forgotten to mention, however, is that every one of them was *not supposed* to detect any linkage. If direct gene analyses have suggested involvement of individual genes only in a few percent of cases, linkage analyses will have no power to detect this effect. In other words: in complex traits, one can confirm linkage by association studies, but usually not association by linkage. Linkage studies are generally hampered by low numbers, in both typed markers and study subjects categories. Genome-wide linkage scans, as they are carried out now, can reliably detect only loci with a substantial involvement in a disorder. Not surprisingly, when a representative set of 101 full-genome linkage studies was analyzed, only 2% of findings could be replicated by an independent study [37]. We do not want to discourage the use of linkage analyses for complex traits, since they will, no doubt, provide us with plethora of information in the future. However, we emphasize that one should fully acknowledge the strong and weak sides of this powerful technique in order to apply it in proper situations. For example, linkage studies have great potential in isolated populations where the number of genes involved in even a complex disorder is usually limited. To increase the power of this method for genome-wide searches in heterogeneous admixed populations, better study design with careful selection of study subjects, and substantial increase in numbers of informative markers resulting in denser maps will be necessary.

Case-control association studies have been considered a good alternative to linkage-based methods. Currently, genome-wide association studies utilizing SNPs are even less feasible than linkage studies, but candidate gene-based studies have appeared promising. They can achieve much better statistical power even on relatively small sample sizes; however, they are also prone to spurious results stemming from population stratification and are limited to only a few selected loci/genes. Fortunately, there are relatively easy solutions to both of these problems. Population stratification can be gauged by genotyping a random set of SNPs or STRs in the study population to assess the matching of patient and control cohorts [38]. Limited numbers of genes and typed markers can be increased significantly by utilizing the most comprehensive genotyping methods available, such as microarrays. Finding an association of rare variants with modest effect with a complex trait is currently feasible only with case-control association studies. New, high-throughput genotyping methods in combination with the currently developed haplotype map of the entire human genome, which includes both microsatellites and SNPs, will facilitate genome-wide association studies and will enhance both linkage-based and LD-based studies.

Conclusions

In summary, the current range of applicable methods does not allow definitive elimination of genes with rare variants from the candidate gene pool. Barring serious deficiencies in mutation detection, variations in these genes, most likely in combinations (as in genotype), may still play a role in AMD. Success in determining the genetic basis of AMD depends on a combination of the following conditions: (1) development of cheap, reliable, high-throughput genotyping methods; (2) substantial increase in the size of study populations and their precise phenotypic characterization (including separation into subclasses, or endophenotypes), and (3) better definition of the pool of candidate genes associated with macular degeneration, including elucidation of their role in normal retinal function. Determining genotypes predisposing to AMD looks feasible already in the near future. This would allow application of molecular diagnostic methods to identify subjects at risk. However, the most anticipated goal – development of genetic research-based treatments for AMD which would apply to a substantial fraction of patients – may prove much more elusive.

References

1 Vingerling JR, Klaver CCW, Hofman A, de Jong PTVM: Epidemiology of age-related maculopathy. Epidemiol Rev 1995;17:347–360.

2 Klaver CCW, Wolfs RCW, Assink JJM, van Duijn CM, Hofman A, de Jong PTVM: Genetic risk of age-related maculopathy. Population-based familial aggregation study. Arch Ophthalmol 1998;116:1646–1651.

3 Hammond CJ, Webster AR, Snieder H, Bird AC, Gilbert CE, Spector TD: Genetic influence on early age-related maculopathy: A twin study. Ophthalmology 2002;109:730–736.

4 Petrukhin K, Koisti MJ, Bakall B, Li W, Xie G, Marknell T, Sandgren O, Forsman K, Holmgren G, Andreasson S, Vujic M, Bergen AA, McGarty-Dugan V, Figueroa D, Austin CP, Metzker ML, Caskey CT, Wadelius C: Identification of the gene responsible for Best macular dystrophy. Nat Genet 1998;19:241–247.

5 Sun H, Tsunenari T, Yau KW, Nathans J: The vitelliform macular dystrophy protein defines a new family of chloride channels. Proc Natl Acad Sci USA 2002;99:4008–4013.

6 Kramer F, White K, Pauleikhoff D, Gehrig A, Passmore L, Rivera A, Rudolph G, Kellner U, Andrassi M, Lorenz B, Rohrschneider K, Blankenagel A, Jurklies B, Schilling H, Schutt F, Holz FG, Weber BH: Mutations in the VMD2 gene are associated with juvenile-onset vitelliform macular dystrophy (Best disease) and adult vitelliform macular dystrophy but not age-related macular degeneration. Eur J Hum Genet 2000;8:286–292.

7 Seddon JM, Afshari MA, Sharma S, Bernstein PS, Chong S, Hutchinson A, Petrukhin K, Allikmets R: Assessment of mutations in the Best macular dystrophy (VMD2) gene in patients with adult-onset foveomacular vitelliform dystrophy, age-related maculopathy, and bull's-eye maculopathy. Ophthalmology 2001;108:2060–2067.

8 Lotery AJ, Munier FL, Fishman GA, Weleber RG, Jacobson SG, Affatigato LM, Nichols BE, Schorderet DF, Sheffield VC, Stone EM: Allelic variation in the VMD2 gene in best disease and age-related macular degeneration. Invest Ophthalmol Vis Sci 2000;41:1291–1296.

9 Stone EM, Lotery AJ, Munier FL, Heon E, Piguet B, Guymer RH, Vandenburgh K, Cousin P, Nishimura D, Swiderski RE, Silvestri G, Mackey DA, Hageman GS, Bird AC, Sheffield VC, Schorderet DF: A single *EFEMP1* mutation associated with both malattia leventinese and Doyne honeycomb retinal dystrophy. Nat Genet 1999;22:199–202.

10 Matsumoto M, Traboulsi EI: Dominant radial drusen and ARG345TRP EFEMP1 Mutation. Am J Ophthalmol 2001;810–812.

11 Knott V, Downing AK, Cardy CM, Handford P: Calcium binding properties of an epidermal growth factor-like domain pair from human fibrillin-1. J Mol Biol 1996;255:22–27.
12 Tarttelin EE, Gregory-Evans CY, Bird AC, Weleber RG, Klein ML, Blackburn J, Gregory-Evans K: Molecular genetic heterogeneity in autosomal dominant drusen. J Med Genet 2001;38:381–384.
13 Sauer CG, White K, Kellner U, Rudolph G, Jurklies B, Pauleikhoff D, Weber BH: EFEMP1 is not associated with sporadic early-onset drusen. Ophthalmic Genet 2001;22:27–34.
14 Weber BHF, Vogt G, Pruett RC, Stohr H, Felbor U: Mutations in the tissue inhibitor of metalloproteinases-3 (TIMP3) in patients with Sorsby's fundus dystrophy. Nat Genet 1994;8:352–355.
15 Yeow KM, Kishnani NS, Hutton M, Hawkes SP, Murphy G, Edwards DR: Sorsby's fundus dystrophy tissue inhibitor of metalloproteinase inhibitory activities, but affect cell adhesion to the extracellular matrix. Matrix Biol 2002;21:75–88.
16 Assink JJ, de Backer E, ten Brink JB, Kohno T, de Jong PT, Bergen AA, Meire F: Sorsby fundus dystrophy without a mutation in the TIMP-3 gene. Br J Ophthalmol 2000;84:682–686.
17 Ayyagari R, Griesinger IB, Bingham E, Lark KK, Moroi SE, Sieving PA: Autosomal dominant hemorrhagic macular dystrophy not associated with the TIMP3 gene. Arch Ophthalmol 2000;118:85–92.
18 De La Paz MA, Pericak-Vance MA, Lennon F, Haines JL, Seddon JM: Exclusion of TIMP3 as a candidate locus in age-related macular degeneration. Invest Ophthalmol Vis Sci 1997;38: 1060–1065.
19 Felbor U, Doepner D, Schneider U, Zrenner E, Weber BHF: Evaluation of the gene encoding the tissue inhibitor of metalloproteinases-3 in various maculopathies. Invest Ophthalmol Vis Sci 1997;38:1054–1059.
20 Stone EM, Sheffield VC, Hageman GS: Molecular genetics of age-related macular degeneration. Hum Mol Genet 2001;10:2285–2292.
21 Zhang K, Kniazeva M, Han M, Li W, Yu Z, Yang Z, Li Y, Metzker ML, Allikmets R, Zack DJ, Kakuk LE, Lagali PS, Wong PW, MacDonald IM, Sieving PA, Figueroa DJ, Austin CP, Gould RJ, Ayyagari R, Petrukhin K: A 5-bp deletion in *ELOVL4* is associated with two related forms of autosomal dominant macular dystrophy. Nat Genet 2001;27:89–93.
22 Li Y, Marcos I, Borrego S, Yu Z, Zhang K, Antino G: Evaluation of the ELOVL4 gene in families with retinitis pigmentosa linked to the RP25 locus. J Med Genet 2001;38:478–480.
23 Ayyagari R, Zhang K, Hutchinson A, Yu Z, Swaroop A, Kakuk LE, Seddon JM, Bernstein PS, Lewis RA, Tammur J, Yang Z, Li Y, Zhang H, Yashar BM, Liu J, Petrukhin K, Sieving PA, Allikmets R: Evaluation of the ELOVL4 gene in patients with age-related macular degeneration. Ophthalmic Genet 2001;22:233–239.
24 Kaplan J, Gerber S, Larget-Piet D, Rozet JM, Dollfus H, Dufier JL, Odent S, Postel-Vinay A, Janin N, Briard ML: A gene for Stargardt's disease (fundus flavimaculatus) maps to the short arm of chromosome 1. Nat Genet 1993;5:308–311.
25 Anderson KL, Baird L, Lewis RA, Chinault AC, Otterud B, Leppert M, Lupski JR: A YAC contig encompassing the recessive Stargardt disease gene (STGD) on chromosome 1p. Am J Hum Genet 1995;57:1351–1363.
26 Allikmets R, Singh N, Sun H, Shroyer NF, Hutchinson A, Chidambaram A, Gerrard B, Baird L, Stauffer D, Peiffer A, Rattner A, Smallwood P, Li Y, Anderson KL, Lewis RA, Nathans J, Leppert M, Dean M, Lupski JR: A photoreceptor cell-specific ATP-binding transporter gene (ABCR) is mutated in recessive Stargardt macular dystrophy. Nat Genet 1997;15:236–246.
27 Shroyer NF, Lewis RA, Lupski JR: Complex inheritance of ABCR mutations in Stargardt disease: Linkage disequilibrium, complex alleles, and pseudodominance. Hum Genet 2000;106:244–248.
28 Allikmets R: Simple and complex ABCR: Genetic predisposition to retinal disease. Am J Hum Genet 2000;67:793–799.
29 Allikmets R and the International ABCR Screening Consortium: Further evidence for an association of ABCR alleles with age-related macular degeneration. Am J Hum Genet 2000;67: 487–491.
30 Mata NL, Tzekov RT, Liu X, Weng J, Birch DG, Travis GH: Delayed dark-adaptation and lipofuscin accumulation in abcr+/– mice: Implications for involvement of ABCR in age-related macular degeneration. Invest Ophthalmol Vis Sci 2001;42:1685–1690.
31 Reich DE, Lander ES: On the allelic spectrum of human disease. Trends Genet 2001;17:502–510.

32 Strittmatter WJ: Apolipoprotein E and Alzheimer's disease. Ann NY Acad Sci 2000;924:91–92.
33 Weiss KM, Clark AG: Linkage disequilibrium and the mapping of complex human traits. Trends Genet 2002;18:19–24.
34 Lander E, Kruglyak L: Genetic dissection of complex traits: Guidelines for interpreting and reporting linkage results. Nat Genet 1995;11:241–247.
35 Weeks DE, Conley YP, Tsai HJ, Mah TS, Rosenfeld PJ, Paul TO, Eller AW, Morse LS, Dailey JP, Ferrell RE, Gorin MB: Age-related maculopathy: an expanded genome-wide scan with evidence of susceptibility loci within the 1q31 and 17q25 regions. Am J Ophthalmol 2001;132:682–692.
36 Weeks DE, Conley YP, Mah TS, Paul TO, Morse L, Ngo-Chang J, Dailey JP, Ferrell RE, Gorin MB: A full genome scan for age-related maculopathy. Hum Mol Genet 2000;9:1329–1349.
37 Altmuller J, Palmer LJ, Fischer G, Scherb H, Wjst M: Genomewide scans of complex human diseases: True linkage is hard to find. Am J Hum Genet 2001;69:936–950.
38 Pritchard JK: Are rare variants responsible for susceptibility to complex diseases? Am J Hum Genet 2001;69:124–137.

Rando Allikmets, PhD
Department of Ophthalmology, Columbia University,
Eye Institute Research, Rm 715, 630 West 168th Street, New York, NY 10032 (USA)
Tel. +1 212 305 8989, Fax +1 212 305 7014, E-Mail rla22@columbia.edu

Wissinger B, Kohl S, Langenbeck U (eds): Genetics in Ophthalmology.
Dev Ophthalmol. Basel, Karger, 2003, vol 37, pp 170–187

Genetics of Color Vision Deficiencies

Samir S. Deeb[a], *Susanne Kohl*[b]

[a] Departments of Medicine and Genome Sciences, University of Washington, Seattle Wash., USA and
[b] Molecular Genetics Laboratory University Eye Hospital, Tübingen, Germany

Abstract

The normal X-chromosome-linked color vision gene array is composed of a single red pigment gene followed by one or more green pigment genes. The high degree of homology between these genes predisposed them to unequal recombination, leading to gene deletions or the formation of red-green hybrid genes that explain the majority of the common red-green color vision deficiencies. Gene expression studies suggest that only the two most proximal genes of the array are expressed in the retina. The severity of the color vision defect is roughly related to the difference in absorption maxima of the photopigments encoded by the first two genes of the array. A single amino acid polymorphism (Ser180Ala) in the red pigment accounts for the subtle difference in normal color vision and influences the severity of color vision deficiency. Blue cone monochromacy is a rare disorder that involves absence of red and green cone function. It is caused either by deletion of a critical region that regulates expression of the red/green gene array, or by mutations that inactivate the red and green pigment genes. Total color blindness is another rare disease that involves complete absence of all cone function. A number of mutations in the genes encoding the cone-specific α- and β-subunits of the cation channel and the α-subunit of transducin have been implicated in this disorder.

Introduction

The human retina contains four classes of photoreceptors: rods, which are used for vision in dim light, and three classes of cone, which are used for vision in bright light and for color vision. Normal color vision is mediated by the three classes of cone photoreceptors (trichromatic color vision), the blue (short-wave sensitive), the green (middle-wave sensitive) and red (long-wave sensitive). All Old World primates and some New World primates have trichromatic color vision. The majority of the other mammals have dichromatic color vision, relying on two classes of cones for limited color discrimination capacity. The introduction of trichromacy into the Old World lineage occurred some 40 million years ago as a result of duplication of the ancestral middle-long photopigment gene on the X-chromosome, followed by divergence into the red and green pigment genes. Trichromacy in Old World primates must have provided some selective advantage since the addition of the new red-green color vision system increased the number of colors that can be discriminated from about 10,000 to about 1 million [1, 2]

There is wide variation in both normal and defective color vision among humans. The inherited forms of color vision deficiencies are classified into four main categories red (protan) or green (deutan) cone deficiencies, blue cone deficiency (tritan), red and green cone deficiency called blue cone monochromacy (BCM), and complete cone deficiency (achromatopsia). The severity of color vision defects varies widely and forms the basis for sub-classification of the defects. The red-green deficiencies, which are inherited as X-chromosome-linked recessive traits, are by far the most common, reaching an incidence of as high as 8% among males of northern European extraction and ~5% among other ethnic groups. The other forms are quite rare.

Inherited forms of color vision deficiencies have well-defined and stable characteristics, and are distinct from the acquired and progressive forms of color vision defects that are encountered during the course of certain ophthalmic diseases such as cone dystrophy. This chapter focuses on describing recent advances in our understanding of the molecular mechanisms underlying the inherited color vision deficiencies. For further detailed information, the reader is referred to three recent reviews on this topic [3–5].

Color Vision Testing

Many types of tests of color vision have been designed, some that are simple and rapid for use in mass screening or in the clinical setting, and others that are highly sophisticated and accurate for use in the laboratory setting. The reader is referred to a comprehensive review of color vision testing for more detailed description [6]. Two main categories of tests will be briefly described: the plate tests and anomaloscopy.

Plate Tests

The Ishihara test (24 plates in the standard test) is the most universally used test to screen for inherited red-green color vision defects. It incorporates designs containing colored numbers or figures against a colored background,

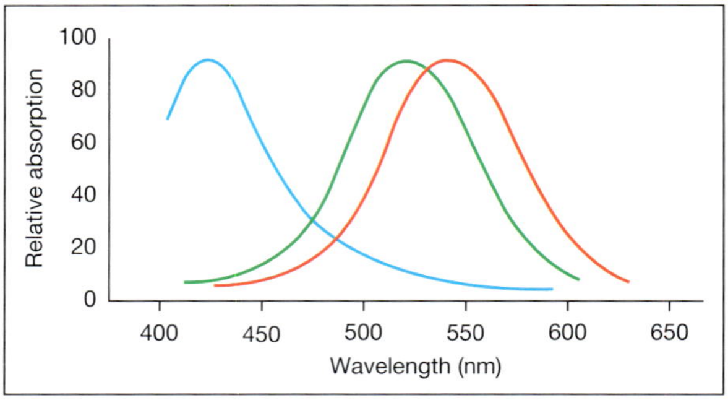

Fig. 1. Absorption spectra of human cones. Shown are the approximate spectra of the blue, green and red classes of cone. The difference between the wavelength of maximum absorption of the red and green cones is approximately 30 nm.

some of which can be seen by individuals with normal color vision, others by those with defective color vision. They also contain diagnostic plates that allow differentiation of protan from deutan observers and of dichromats from mild (but not from severe) anomalous trichromats. A special four-color plate test is used to distinguish BCM from achromatopsia [7].

Anomaloscopy

The Rayleigh match, as measured by anomaloscopy (Nagel type I anomaloscope) has been used widely in clinical and laboratory situations for the accurate diagnosis of red-green color vision defects. It is based on matching the color of a yellow (589 nm) primary light to a juxtaposed mixture of red (679 nm) and green (544 nm) primary lights. Normal observers accept a much narrower range of mixtures of red and green lights than color-defective individuals. Dichromats such as protanopic and deuteranopic subjects will match yellow with any and all ratios of red and green, including red and green alone simply by adjusting the intensity of light. Among protanomalous and deuteranomalous subjects, the wider the match range, the more severe is the defect.

Photoreceptors and Photopigments

Recent advances in the molecular biology of the human cone visual pigments and the genes that encode them have led to elucidation of the molecular basis of inherited differences in color vision. Normal color vision is mediated by three classes of cone photoreceptors that contain blue (short-wave sensitive), green (middle-wave sensitive) or red (long-wave sensitive) photopigments with absorption maxima (λ_{max}) at around 420, 530 and 560 nm, respectively (fig. 1).

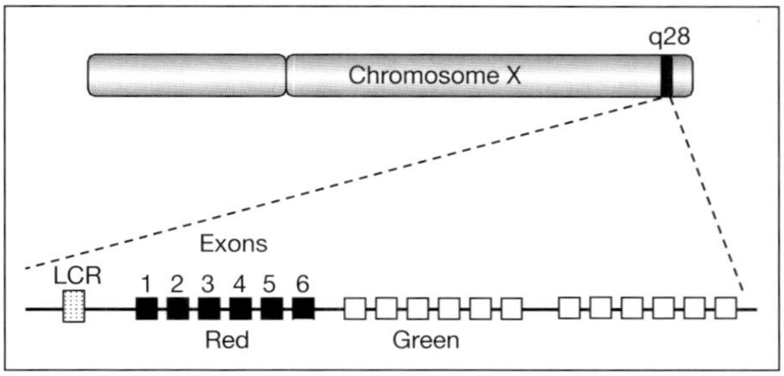

Fig. 2. The visual pigment gene array on the X-chromosome. Diagram of the X-chromosome indicating the position of the red and green pigment gene array. The gene array is composed of a locus control region (LCR) that is essential for expression of the genes of the locus, followed by one red pigment gene (~3.6 kb from the LCR) and one or more green pigment genes. Filled and open boxes represent exons of the red and green pigment genes, respectively. The genes are approximately 15 kb in length and the intergenic regions are approximately 25 kb in length (diagram not to scale).

Each cone photoreceptor cell expresses only one of the photopigments. A small (0.3 mm in diameter) centrally located area of the retina, called the fovea, has evolved for high visual acuity. It is densely packed with red and green cones (~10,000). Rods and blue cones are found outside the fovea. The photopigment spectra have wide regions of overlap. This is critical for color vision, since perception of color in the brain results from comparison between signal outputs from the three classes of cone photoreceptors. For example the ratio of outputs from the red and green cones for monochromatic light of a wavelength of 600 nm would be about 2.5, with very little if any contribution from blue cones. As shown in figure 1, this ratio changes as a function of the wavelength of light.

The visual pigments belong to the heptahelical transmembrane receptor family that includes the olfactory receptors. The red and green photopigments are very similar in amino acid sequence, differing at only 15 positions. However, differences at only three amino acids (at positions 180, 277 and 285) account for 28 of the 30-nm difference in λ_{max} between these two pigments [8, 9].

Genes Encoding the Cone Photopigments

The genes encoding the red and green photopigments are arranged in a head-to-tail tandem array on the X-chromosome (Xq28). The arrays are comprised of a single red pigment gene (six exons) and one or more green pigment genes (fig. 2). The blue pigment gene is located on chromosome 7. Approximately 25% of male Caucasians have a single green pigment gene,

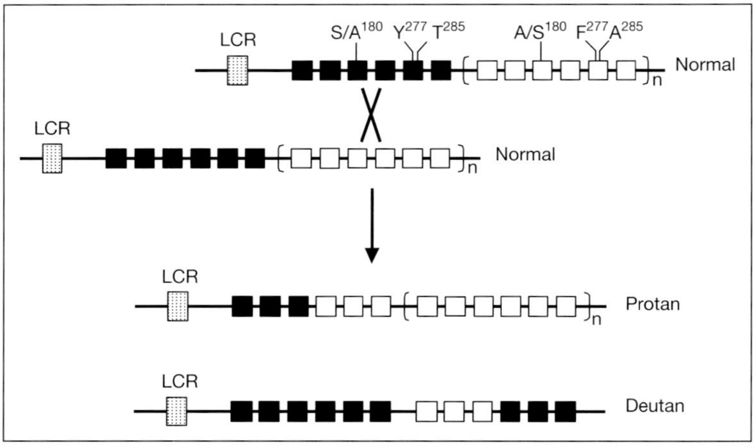

Fig. 3. Unequal homologous recombination (crossover) between the red and green pigment genes to form hybrid genes. Filled and open rectangles represent red and green exons, respectively. Recombination between the red and green pigment genes within intron 3 gives rise to a red-green hybrid gene that is typically found in protans, and a green-red hybrid gene that is typically found in deutans. The three amino acid residues that contribute the majority of the difference in absorption maxima (λ_{max}) between the encoded red and green pigments are indicated above the array. Position 180 is polymorphic (Ala or Ser) in the general population. The locus control region (LCR) is essential for expression of all the pigment genes of an array.

50% have two, while the rest have three or more green pigment genes. The expression of the genes of the array is controlled by a highly conserved sequence of DNA, referred to as the locus control region (LCR), located approximately 3.5 kb upstream of the red pigment gene [10]. Deletion of the LCR was shown to be associated with loss of expression of all the genes of the array [11], resulting in BCM (see below). The LCR was also shown to play a critical role in ensuring mutually exclusive expression of the genes of the array in single cones (i.e. a single cone expresses only one of the pigment genes) [12]. A proposed mechanism by which this occurs is described below. Only the first two genes of an array are sufficiently expressed in the retina to influence the color vision phenotype. Proximity to the LCR appears to be a significant factor in expression of the genes of this locus [12].

The high degree of homology between the red and green pigment genes (including introns and intergenic sequences) has predisposed the locus to unequal homologous recombination. These illegitimate events result in a change in the number of green pigment genes (including their total elimination) or in the formation of red/green hybrid genes (fig. 3) that are often associated with color vision deficiencies. These illegitimate recombination events are much more frequent than point mutations. Therefore, the high sequence homology between the red and green pigment genes and their juxtaposition on the X-chromosome account for the relatively high frequency of red/green color vision defects.

Variation in Normal Color Vision

Subtle variation in color perception in the red-green region of the spectrum was observed among humans with normal color vision. This variation was suggested to be due to a small variation in the λ_{max} of red or green visual pigments [13]. Molecular analysis of the visual pigment genes revealed a common polymorphism (Ser180Ala) in the red pigment. The frequency of the Ala form of the red pigment among Caucasians is approximately 0.40 [14]. In the green pigment, approximately 86% of X-chromosomes encode Ala and 15% have Ser at position 180. This polymorphism was detected among a variety of populations including those of African and Asian extraction. Interestingly, males who carry a red pigment with Ser at position 180 were shown, by color-matching experiments, to be more sensitive to red light than those who carried a red pigment with Ala [14]. The λ_{max} of a red photopigment with Ser at position 180 was subsequently shown to be ~4 nm longer than that with Ala [8]. Thus, the Ser180Ala polymorphism has generated two forms of each of the red and green visual pigments. As will be discussed below, this polymorphism also plays an important role in determining the severity of red-green color vision deficiencies.

Due to random X-chromosome inactivation, the retinae of females who carry both the Ser and Ala forms of the red pigment (expressed in different cones) contain a total of four classes of cone (red^{Ser}, red^{Ala}, green and blue), and have the potential for tetrachromatic color vision [15].

Molecular Genetics of Color Vision Defects

Classes of color vision deficiency include: (1) the common red-green defects that include the protan type caused either by lack of red cones (protanopia) or by replacement of red cones with ones that contain anomalous pigments (protanomaly), and the deutan type caused either by lack of green cones or by replacement of green cones with ones that contain anomalous pigments (deuteranomaly); (2) BCM due to loss of both red and green cones; (3) the blue-yellow or tritan color vision due to nonfunctional blue cones, and (4) achromatopsia or complete color blindness due to loss of function of all three classes of cone.

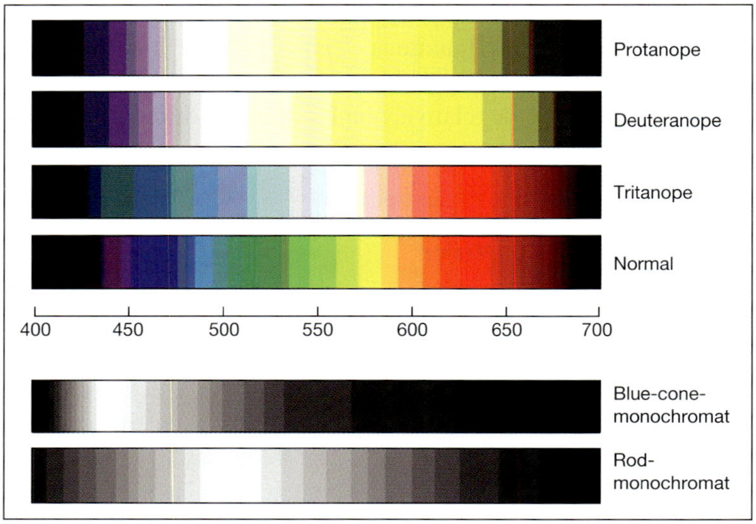

Fig. 4. The visible spectrum as it appears to observers with normal and with various classes of defective color vision [from 3, with permission].

An approximation of how the colors of the rainbow appear to individuals with normal color vision and to those with various classes of defective color vision is shown in figure 4.

1. The Common Red-Green Color Vision Defects

Cloning of the genes that encode the blue, red and green photoreceptor pigments by Nathans et al. [16] opened the door to defining the molecular bases of color vision deficiencies. These investigators were the first to show that color vision defects are associated with gene deletions and the formation of red/green hybrid genes [17]. The following summarizes our present knowledge of the molecular bases of red-green color vision defects.

Protan color vision defects: Protan color vision defects result from defects in the red photopigment and red cones. Subjects with the severe form, called protanopia (dichromatic color vision), have no red photopigment. Subjects with the milder form, called protanomaly (anomalous trichromatic color vision), have an anomalous red (green-like) photopigment with a λ_{max} value that differs from that of the green by 2–7 nm (fig. 5). The lower the degree of spectral separation between the anomalous red and the normal green pigments, the more severe is the color vision defect. The genetic basis of this class of defects is the formation of a red-green hybrid gene in the first position of the array due to unequal crossing over between the red and green pigment genes as illustrated

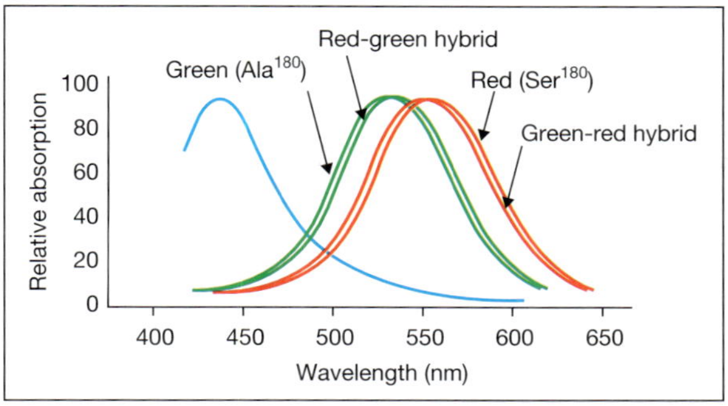

Fig. 5. The absorption spectral of the green-like and red-like cones. Protanomalous subjects have the blue, the green and the green-like cones. Deuteranomalous subjects have the blue, red and red-like cones.

in figure 3. The gene arrays of protans are illustrated in figure 6a. Note that if the red-green hybrid and normal green pigment genes encode pigments with identical λ_{max} value, the color vision defect is protanopia; if they encode pigments with different λ_{max} values, the defect is protanomaly. Note that the Ser180Ala polymorphism plays an important role in the spectral separation between the red-green hybrid and normal green pigments and, therefore, in the severity of the color vision defect.

Deutan color vision defects: Deutan color vision results from defects in the green photopigment. Subjects who lack functional green cones have the sever form, called deuteranopia (dichromatic color vision). Those with the milder defect, called deuteranomaly (anomalous trichromacy), have an anomalous green pigment (red-like) with a λ_{max} that differs from that of the normal red by 2–7 nm instead of 30 nm (fig. 5).

The gene arrays of deutans are illustrated in figure 6b. Deuteranopia results either from the deletion of the green pigment gene(s) or the formation of green-red hybrid genes (by unequal crossing over, see fig. 3) that encode a pigment of identical λ_{max} to the normal red pigment. Deuteranomaly results if the green-red hybrid gene encodes a pigment that differs in λ_{max} from the normal red by 2–7 nm. As in protans, the Ser180Ala polymorphism plays a role in spectral separation between the normal and hybrid pigments and, therefore, in the severity of the color vision defect. Green-red hybrid genes are a common cause of deutan color vision defects. A rare cause of such defects is the inactivating point mutation C203R [18, 19].

Surprisingly, a large proportion of deutans (both deuteranopes and deuteranomalous subjects) have, in addition to a normal red and a green-red hybrid gene, one or more normal green pigment genes. Normal color vision would be

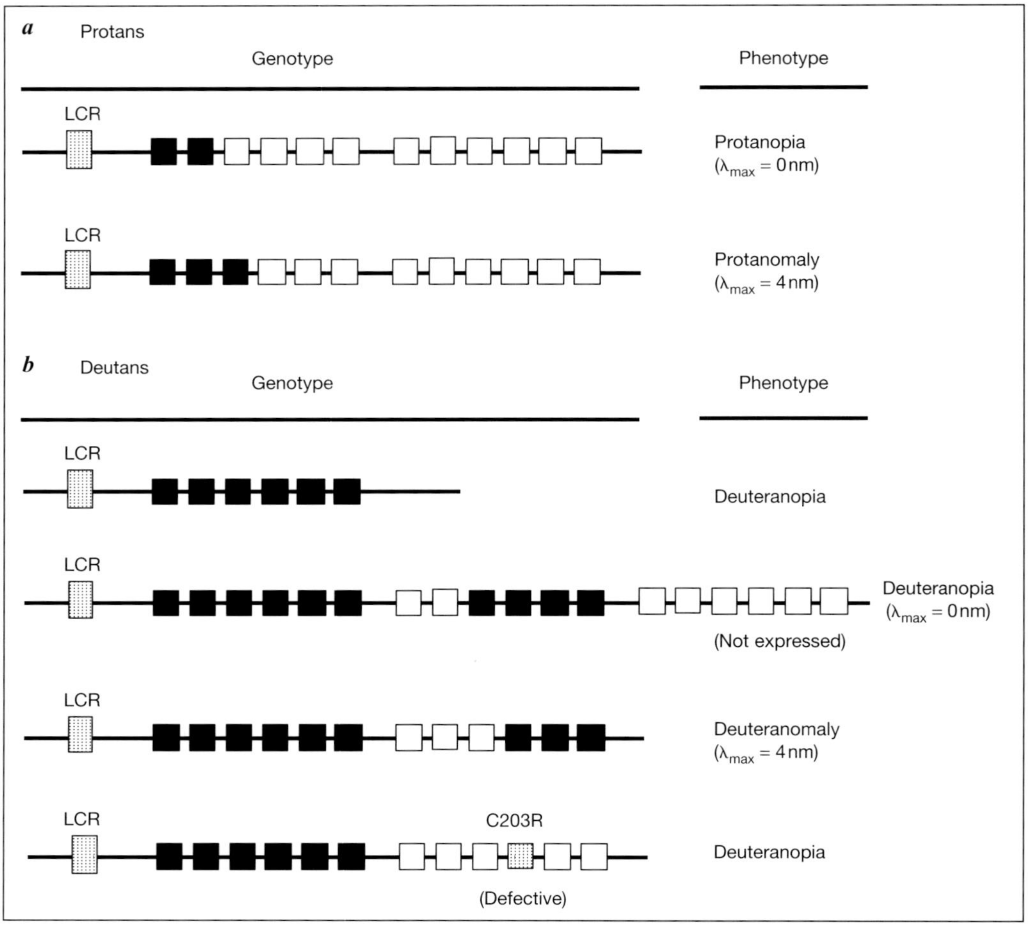

Fig. 6. Genotype-phenotype relationships among males with protan and deutan color vision defects. Shown are examples of gene arrays found among males with protan (***a***) and deutan (***b***) color vision. Filled and open squares represent exons of the red and green pigment genes, respectively. Fusion points in intron 1, 2, 3, or 4 were observed. The red-green hybrid genes that occupy the first position in the array cause protanopia (dichromacy) if the first two genes encode photopigments with identical λ_{max} and cause deuteranomaly if they encode photopigments that differ in λ_{max} by 2–7 nm (see ***a***). Deuteranopia (dichromacy) results from deletion of the green pigment gene. Green-red hybrid genes that occupy the second position in the array cause deutan color vision. As in protans, deuteranopia results if the first two genes encode identical pigments, and deuteranomaly is observed if the two encoded pigment differ by 2–7 nm (see ***b***). Genes that occupy third or more distal positions are not expressed in the retina and do not influence the color vision phenotype. Shown also is the Cys203Arg mutation as a less common cause of deutan color vision. The locus control region (LCR) is essential for expression of all the genes in the array.

expected if all of the genes of the array are expressed. In addition, green-red hybrid genes that are similar to those associated with deutan color vision, are also found in 4–8% of males with normal color vision [19–21]. To explain these findings, it was hypothesized that only the first two genes of the array are expressed in the retina and, therefore, participate in determining the color vision phenotype. It was proposed that during cone photoreceptor differentiation, the LCR (the major switch that controls expression of the red and green pigment genes) couples to and activates expression of either the red pigment gene (the first gene in the array) to form red cone photoreceptors [10, 22]. Alternatively, the LCR couples to and activates the second gene in the array to form either a normal green cone or a cone with a green-red hybrid pigment. More distal genes are too far removed from the LCR to be activated by the LCR. Thus, the hybrid genes in deuteranomalous subjects occupy the second position in the array and are expressed instead of the normal green pigment genes. The green red hybrid genes found in a small percentage of subjects with normal color vision occupy third or more distal positions in the array and are not expressed. There is good experimental evidence that only the first two genes of a visual pigment array are expressed in the retina and participate in determining the color vision phenotype [19, 22–24].

2. Blue Cone Monochromacy

BCM (MIM303700), also known as X-chromosome-linked incomplete achromatopsia, is a rare X-linked ocular disorder, characterized by poor visual acuity, infantile nystagmus (which diminishes with age), and photodysphobia, together with severely reduced color discrimination capacity. It is sometimes associated with progressive macular atrophy. Subjects with BCM have no functional red and green cones, but preserved blue cones and rods. Under photopic conditions, BCM individuals experience total colorblindness, while at intermediate light levels, interactions between rod and blue cone signals allows for crude hue discrimination.

Deletions encompassing the LCR which is required for expression of the red and green pigment genes, as well as point mutations that inactivate the red and green pigments have been implicated in causing BCM [25, 26]. The most common point mutation in BCM is the C203R.

3. Tritan or Blue-Yellow Color Vision Deficiency

Tritan or blue-yellow color vision deficiency is due to defective blue cones and is characterized by blue-yellow color confusion. It is a rare (<1/1,000) autosomal dominant trait with severe (tritanopia) and mild (tritanomaly) forms. The following mutations in the blue pigment gene, located on chromosome 7, have been implicated in causing tritanopia: Gly79Arg, Ser214Pro and

Pro264Ser [27, 28]. These mutations are located in the transmembrane domain of the opsin and are believed to disrupt the structure or stability of the protein. There is evidence for incomplete penetrance of this disorder. The diagnosis of tritan defects is not simple. The most frequently used test is based on the Moreland equation, in which an observer is asked to match a mixture of lights at 436 nm (indigo) and 490 nm (green) to a cyan standard (fixed ratio of 480 and 580 nm) light.

4. Achromatopsia (Total Colorblindness/Rod Monochromacy/ Complete Achromatopsia)

Achromatopsia, also referred to as total colorblindness or rod monochromacy, is an autosomal recessive congenital and stationary ocular disorder with a prevalence of 1 in 30,000. Clinically it is characterized by severe photophobia under daylight conditions and nystagmus, both symptoms become evident within the first months after birth. Visual acuity is strongly reduced to <0.2 and color discrimination is impossible. In ERG recordings, rod function is normal, but cone function is absent or strongly reduced. It is a genetic dysfunction of all three types of cones.

In the past few years, two loci for complete achromatopsia, *ACHM2* (MIM216900) and *ACHM3* (MIM262300), have been assigned by linkage analysis to chromosome 2q11 and 8q21, respectively [29–32]. Subsequently, mutations in the *CNGA3* (MIM600053) and the *CNGB3* genes (MIM605080) were shown to segregate with the *ACHM2* [33] and the ACHM3 locus [34, 35], respectively, in families with complete achromatopsia. Mutations in the *CNGA3* gene account for approximately 20–30%, and mutations in the *CNGB3* gene for 40–50% of all achromatopsia patients. This clearly shows that the *ACHM3* locus on chromosome 8q21 is the major locus for achromatopsia. However, these results also provided evidence for further genetic heterogeneity in this rare disorder.

Recently, mutations in the *GNAT2* gene (MIM139340) on chromosome 1p13 were shown to account for a small percentage (~2%) of this rare disorder [36, 37]. *GNAT2* encodes the cone-specific α-subunit of transducin, a heterotrimeric G-protein that couples to the cone visual pigments. Excited pigment molecules induce the exchange of GDP to GTP at the guanosine-binding site of the transducin α-subunit and its subsequent release from the inhibitory β/γ-subunits. The activated GTP transducin then binds and activates a phosphodiesterase that hydrolyzes cGMP and effectively reduces its intracellular concentration. This results in closure of the cGMP-gated channels and subsequently in membrane hyperpolarization [38]. *CNGA3* and *CNGB3* encode the channel-forming α- and putative modulatory β-subunits of the heterotetrameric cone photoreceptor cGMP-gated (CNG) channel, respectively.

Transducin thus mediates one of the first steps of the phototransduction cascade, while the cGMP-gated channel represents the final component of the very same pathway. Considering the importance of transducin and the CNG channel in phototransduction, the phenotype of achromatopsia can easily be explained by mutations in the *GNAT2*, the *CNGA3* or the *CNGB3* gene. Furthermore, it implies that the same transducin α-subunit and the same CNG channel is common to all three types of cone photoreceptors.

CNGA3 *Mutations*

The *CNGA3* gene is located on chromosome 2q11. Its large terminal exon encodes about two-thirds of the protein containing all functionally conserved domains [30, 39]. Screening of the *CNGA3* gene in patients affected by achromatopsia resulted in the identification of 46 different mutations [33, 39]. The vast majority are hereby missense mutations (fig. 7a) altering residues that are highly conserved across various species [39]. In vitro expression experiments with selected mutant *CNGA3* cDNA constructs showed that the mutations result in the expression of mutant CNGA3 channels with altered or abolished channel properties.

The mutations in *CNGA3* are mainly confined to the functionally and structurally important central and terminal parts of the CNGA3 polypeptide, including the six transmembrane helices (S1–S6), the ion pore and the cGMP-binding domain (fig. 7a).

Several mutations were found recurrently in different patients. The most prevalent mutations are R283W, F547L, R277C and R436W. Together, these four mutations account for over 40% of all identified mutant *CNGA3* chromosomes. Hereby the R283W mutation was found frequently in achromatopsia patients from Scandinavia and Northern Italy, and all observed R283W chromosomes share a common haplotype, implying a founder effect.

CNGA3 mutations were also detected among individuals with incomplete achromatopsia, and in few cases of cone dystrophy. The considerable variability in the clinical presentation among these patients probably reflects the extent to which cone phototransduction is affected. While in some patients residual cone function can only be detected by psychophysical color vision testing, others have clearly measurable cone ERGs and only slightly impaired non-uniform color vision. Visual acuity in these patients is better than in those with complete achromatopsia, and not all incomplete achromatopsia patients complain about photophobia [39].

Analysis of the *CNGA3* knockout mouse showed complete absence of physiologically measurable cone function, a decrease in the number of cones in the retina and morphological abnormalities in the remaining cones [40]. These results are comparable to those of histological examinations of purported

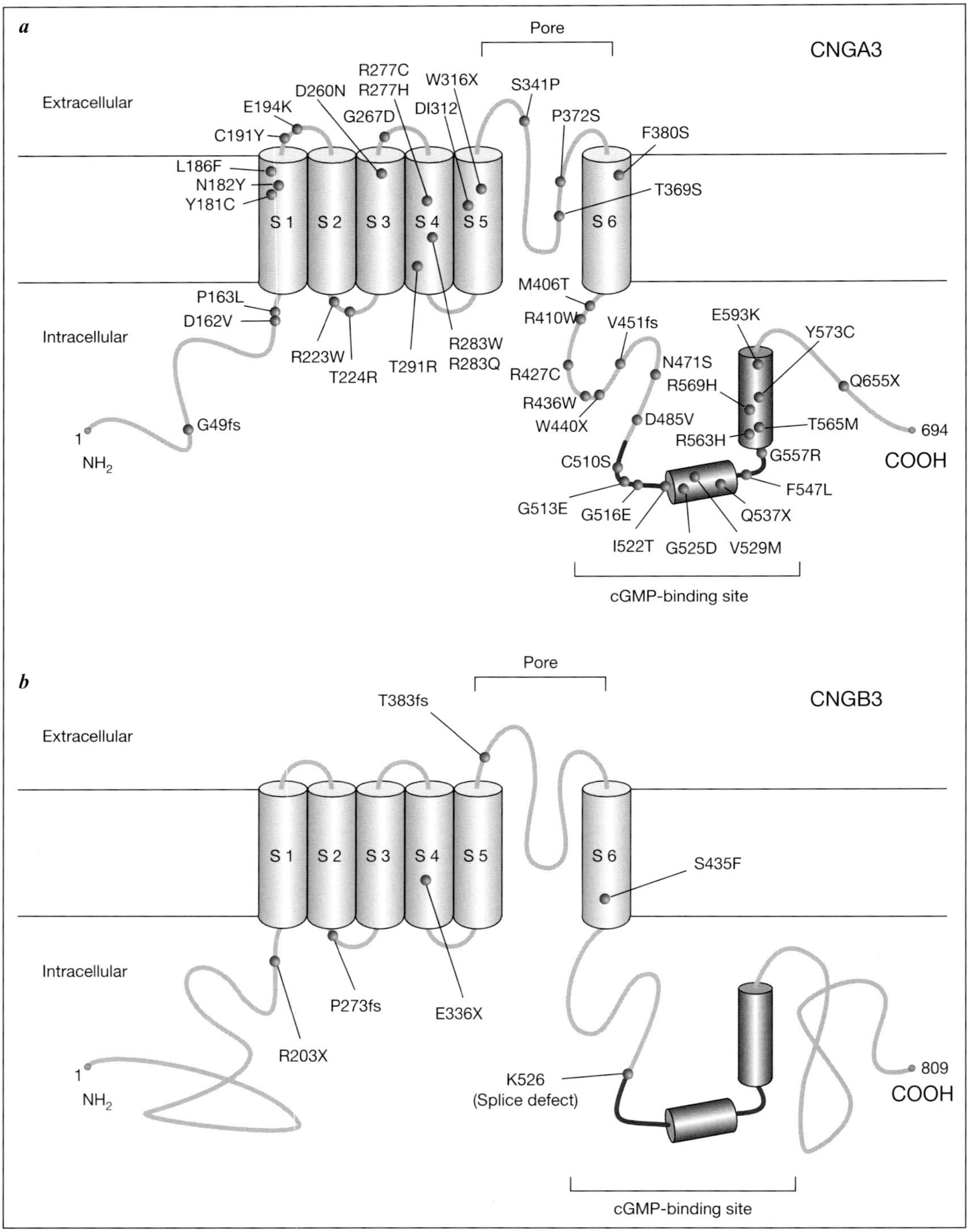
a
Pore
CNGA3
Extracellular
Intracellular
E194K
C191Y
L186F
N182Y
Y181C
D260N
R277C
R277H
G267D
W316X
DI312
S341P
P372S
F380S
T369S
S 1
S 2
S 3
S 4
S 5
S 6
P163L
D162V
R223W
T224R
T291R
R283W
R283Q
M406T
R410W
V451fs
R427C
R436W
W440X
N471S
D485V
E593K
Y573C
R569H
Q655X
T565M
R563H
G557R
694
COOH
C510S
G513E
G516E
I522T
G525D
V529M
Q537X
F547L
cGMP-binding site
G49fs
1
NH2
b
Pore
CNGB3
T383fs
Extracellular
Intracellular
S 1
S 2
S 3
S 4
S 5
S 6
S435F
P273fs
E336X
R203X
K526
(Splice defect)
1
NH2
809
COOH
cGMP-binding site
7

human achromat eyes, although the extent of loss in cone number and their regional distribution varied considerably [41–43]. Yet, variability in histological presentation may either reflect the genetic heterogeneity of achromatopsia and/or inclusion of cases with complete and incomplete achromatopsia or cone dystrophy.

CNGB3 *Mutations*

Thus far, only six different mutations have been observed in the *CNGB3* gene on chromosome 8q21 [35] (fig. 7b). In contrast to the *CNGA3* mutations, the majority of mutations in *CNGB3* result in truncated proteins. Both the nonsense mutations and the deletions, which all create frameshifts, result in premature translation termination well before the cGMP-binding site and most likely abolish any channel function. The putative splice site mutation is also unlikely to result in a functional protein product.

Except for the R203X mutation, all mutations in *CNGB3* have been found recurrently. The 1148delC mutation accounts for over 80% of all *CNGB3* mutant alleles in the Caucasian population, which enables the molecular diagnosis of ~40% of all achromatopsia patients [unpubl. data]. Haplotype reconstruction showed that probably all patients with this mutation carry a common haplotype, implying a founder effect [unpubl. data].

The only missense mutation S435F is the cause of the so-called ‘Pingelapese blindness’ (MIM262300). The *ACHM3* locus has originally been mapped to chromosome 8q21 in Pingelapese pedigrees from a Southwestern Pacific Islander population, in which achromatopsia is highly frequent, affecting nearly 10% of the native population. The very high incidence of achromatopsia among the Pingelapese is thought to be the result of genetic drift following a devastating typhoon in the 17th century that killed most of the inhabitants. This has come to public attention through Oliver Sacks’ book ‘The Island of the Colorblind’ and a BBC TV documentation.

GNAT2 *Mutations*

The human *GNAT2* gene (MIM139340) is located on human chromosome 1p13 [44]. Seven different disease-associated mutations segregating in six independent families with achromatopsia have been observed so far [36, 37].

Fig. 7. CNGA3 and CNGB3 mutations in achromatopsia. Location of the 46 different *CNGA3* (***a***) and the six *CNGB3* (***b***) mutations at the protein level in relation to the putative topology of the human cone cGMP-gated cation channel α- and β-subunit with six transmembrane domains. The membrane-spanning domains S5 and S6 are thought to line the ion-conducting pore, the cGMP-binding site is located at the intracellular carboxyterminus. The locations of the mutations are indicated by gray dots.

In detail these are one nonsense mutation (Q79X) and five small deletion and/or insertion mutations (285–291del7insCTGTAT; 503–504insT; 802–803insTCAA; 842–843insTCAG; 955delA), all of which result in frameshifts. The seventh mutation represents a large intragenic deletion eliminating exon 4 and flanking intronic sequences (IVS3 + 365–IVS4 + 974del). All observed mutations result in mutant polypeptides that lack considerable portions of the genuine carboxyterminus. In the rod photoreceptor paralogue the carboxyterminus contains major interaction sites with the excited rod photopigment [45]. Therefore, the mutations most probably represent functional null alleles of *GNAT2* that prevent the formation of the trimeric G protein complex or its interaction with the excited photopigments.

Conclusions

1. Most of the common red-green color vision defects are caused by either deletion of the green pigment gene or the presence of red-green (protans) or green-red (deutans) hybrid pigment genes. Such gene alterations result from unequal recombination between the red and green pigment genes that are arranged in a head-to-tail tandem array on the X-chromosome.
2. Only the first two genes of the array are expressed in the retina and, therefore, contribute to color vision.
3. The severity of the red-green color vision defects is roughly correlated with the difference in absorption maxima of the pigments encoded by the first two genes of an array.
4. A common amino acid polymorphism (Ala180Ser) in the red pigment plays a role in variation in normal and defective color vision.
5. Mutations in the blue pigment gene, located on chromosome 7, account for the rare tritan color vision defects.
6. Deletions of the LCR or mutations that inactivate the red and green pigment genes cause blue-cone monochromacy.
7. Mutations in the genes encoding the cone-specific α- and β-subunits of the cation channel and the α-subunit of transducin cause total color blindness or complete achromatopsia.

Acknowledgements

Preparation of this chapter was supported by a National Institutes of Health grant number EY08395 to S.D. and a grant of the Deutsche Forschungsmeinschaft (SFB430/A5).

References

1 Mollon JD: Color vision: Opsins and options. Proc Natl Acad Sci USA 1999;96:4743–4745.

2 Regan BC, Julliot C, Simmen B, Vienot F, Charles-Dominique P, Mollon JD: Fruits, foliage and the evolution of primate colour vision. Philos Trans R Soc Lond B Biol Sci 2001;356: 229–283.

3 Sharpe LT, Stockman A, Jagle H, Nathans J: Opsin genes, cone photopigments, color vision and color blindness; in Gegenfurtner KR, Sharpe LT (eds): Color Vision: From Genes to Perception. Cambridge, Cambridge University Press, 1999, pp 3–51.

4 Deeb SS, Hayashi T, Winderickx J, Yamaguchi T: Molecular analysis of human red/green visual pigment gene locus: Relationship to color vision. Methods Enzymol 2000;316:651–670.

5 Motulsky AG, Deeb SS: Color vision and its genetic defects; in Scriver CR, Beaudet AC, Sly WS, Valle D (eds): The Metabolic and Molecular Bases of Inherited Disease, ed 8. New York, McGraw-Hill, 2001, vol IV, pp 5955–5976.

6 Birch J: Diagnosis of Defective Color Vision. Oxford, Oxford University Press, 1993.

7 Berson EL, Sandberg MA, Rosner B, Sullivan PL: Color plates to help identify patients with blue cone monochromatism. Am J Ophthalmol 1983;95:741–747.

8 Merbs SL, Nathans J: Absorption spectra of human cone pigments. Nature 1992;356:433–435.

9 Asenjo AB, Rim J, Oprian DD. Molecular determinants of human red/green color discrimination. Neuron 1994;12:1131–1138.

10 Wang Y, Macke JP, Merbs SL, Zack DJ, Klaunberg B, Bennet J, Gearhart J, Nathans J: A locus control region adjacent to the human red and green visual pigment genes. Neuron 1992; 9:429–440.

11 Nathans J, Davenport CM, Maumenee IH, Sadowski B, Sharpe LT, Lewis RA, Hansen E, Rosenberg T, Schwartz M, Heckenlively JR, Traboulsi E, Klingaman R, Bech-Hansen NT, LaRoche GR, Pagon RA, Murphey WH, Weleber RG: Molecular genetics of human blue cone monochromacy. Science 1989;245:831–838.

12 Smallwood PM, Wang Y, Nathans J: Role of a locus control region in the mutually exclusive expression of human red and green cone pigment genes. Proc Natl Acad Sci USA 2002; 99:1008–1011.

13 Alpern M: Lack of uniformity in colour matching. J Physiol 1979;288:85–105.

14 Winderickx J, Lindsey DT, Sanocki E, Teller DY, Motulsky AG, Deeb SS: Polymorphism in red photopigment underlies variation in colour matching. Nature 1992;356:431–433.

15 Jordan G, Mollon JD: A study of women heterozygous for colour deficiencies. Vision Res 1993;33:1495–1508.

16 Nathans J, Thomas D, Hogness DS: Molecular genetics of human color vision: The genes encoding blue, green and red pigments. Science 1986;232:193–202.

17 Nathans J, Piantanida TP, Eddy RL, Shows TB, Hogness DS: Molecular genetics of inherited variation in human color vision. Science 1986;232:203–210.

18 Winderickx J, Sanocki E, Lindsey DT, Teller DY, Motulsky AG, Deeb SS: Defective colour vision associated with a missense mutation in the human green visual pigment gene. Nat Genet 1992;1:251–256.

19 Jagla WM, Jagle H, Hayashi T, Sharpe LT, Deeb SS: The molecular basis of dichromatic color vision in males with multiple red and green visual pigment genes. Hum Mol Genet 2002; 11:23–32.

20 Deeb SS, Lindsey DT, Hibiya Y, Sanocki E, Winderickx J, Teller DY, Motulsky AG: Genotype-phenotype relationships in human red/green color-vision defects: Molecular and psychophysical studies. Am J Hum Genet 1992;51:687–700.

21 Neitz J, Neitz M, Kainz PM: Visual pigment gene structure and the severity of color vision defects. Science 1996;274:801–804.

22 Winderickx J, Battisti L, Motulsky AG, Deeb SS: Selective expression of human X chromosome-linked green opsin genes. Proc Natl Acad Sci USA 1992;89:9710–9714.

23 Yamaguchi T, Motulsky AG, Deeb SS: Visual pigment gene structure and expression in human retinae. Hum Mol Genet 1997;6:981–990.

24 Hayashi T, Yamaguchi T, Kitahara K, Sharpe LT, Jagle H, Yamade S, Ueyama H, Motulsky AG, Deeb SS: The importance of gene order in expression of the red and green pigment genes and in color vision. Col Res Appl 2001;26:S79–S83.
25 Nathans J, Maumenee IH, Zrenner E, Sadowski B, Sharpe LT, Lewis RA, Hansen E, Rosenberg T, Schwartz M, Heckenlively JR, Traboulsi E, Klingaman R, Bech-Hansen NT, LaRoche GR, Pagon RA, Murphey WH, Weleber RG: Genetic heterogeneity among blue-cone monochromats. Am J Hum Genet 1993;53:987–1000.
26 Ayyagari R, Kakuk LE, Bingham EL, Szczesny JJ, Kemp J,Toda Y, Felius J, Sieving PA: Spectrum of color gene deletions and phenotype in patients with blue cone monochromacy. Hum Genet 2000;107:75–82.
27 Weitz CJ, Went LN, Nathans J: Human tritanopia associated with a third amino acid substitution in the blue-sensitive visual pigment. Am J Hum Genet 1992;51:444–446.
28 Weitz CJ, Miyake Y, Shinzato K, Montag E, Zrenner E, Went LN, Nathans J: Human tritanopia associated with two amino acid substitutions in the blue-sensitive opsin. Am J Hum Genet 1992;50:498–507.
29 Arbour NC, Zlotogora J, Knowlton RG, Merin S, Rosenmann A, Kanis AB, Rokhlina T, Stone EM, Sheffield VC: Homozygosity mapping of achromatopsia to chromosome 2 using DNA pooling. Hum Mol Genet 1997;6:689–694.
30 Wissinger B, Jagle H, Kohl S, Broghammer M, Baumann B, Hanna DB, Hedels C, Apfelstedt-Sylla E, Randazzo G, Jacobson SG, Zrenner E, Sharpe, LT: Human rod monochromacy: Linkage analysis and mapping of a cone photoreceptor expressed candidate gene on chromosome 2q11. Genomics 1998;51:325–331.
31 Winick JD, Blundell ML, Galke BL, Salam AA, Leal SM, Karayiorgou M: Homozygosity mapping of the achromatopsia locus in the Pingelapese. Am J Hum Genet 1999;64:1679–1685.
32 Milunsky A, Huang XL, Milunsky J, DeStefano A, Baldwin CT: A locus for autosomal recessive achromatopsia on human chromosome 8q. Clin Genet 1999;56:82–85.
33 Kohl S, Marx T, Giddings I, Jagle H, Jacobson SG, Apfelstedt-Sylla E, Zrenner E, Sharpe LT, Wissinger B: Total colour blindness is caused by mutations in the gene encoding the α-subunit of the cone photoreceptor cGMP-gated cation channel. Nat Genet 1998;19:257–259.
34 Sundin OH, Yang JM, Li Y, Zhu D, Hurd JN, Mitchell TN, Silva ED, Maumenee IH: Genetic basis of total colour blindness among the Pingelapese islanders. Nat Genet 2000;25:289–293
35 Kohl S, Baumann B, Broghammer M, Jagle H, Sieving P, Kellner U, Spegal R, Anastasi M, Zrenner E, Sharpe LT, Wissinger B: Mutations in the CNGB3 gene encoding the β-subunit of the cone photoreceptor cGMP-gated channel are responsible for achromatopsia (ACHM3) linked to chromosome 8q21. Hum Mol Genet 2000;9:2107–2116.
36 Kohl S, Baumann B, Rosenberg T, Kellner U, Lorenz B, Vadala M, Jacobson SG, Wissinger B: Mutations in the cone photoreceptor G-protein α-subunit gene GNAT2 in patients with achromatopsia. Am J Hum Genet 2002;71:422–425.
37 Aligianis IA, Forshew T, Johnson S, Michaelides M, Johnson CA, Trembath RC, Hunt DM, Moore AT, Maher ER: Mapping of a novel locus for achromatopsia (ACHM4) to 1p and identification of a germline mutation in the α-subunit of cone transducin (GNAT2). J Med Genet 2002;39:656–660.
38 Stryer L. Vision: From photon to perception. Proc Natl Acad Sci USA 1996;93:557–559.
39 Wissinger B, Gamer D, Jagle H, Giorda R, Marx T, Mayer S, Tippmann S, Broghammer M, Jurklies B, Rosenberg T, Jacobson SG, Sener EC, Tatlipinar S, Hoyng CB, Castellan, Bitoun P, Andreasson S, Rudolph G, Kellner U, Lorenz B, Wolff G, Verellen-Dumoulin C, Schwartz M, Cremers FP, Apfelstedt-Sylla E, Zrenner E, Salati R, Sharpe LT, Kohl S: CNGA3 mutations in hereditary cone photoreceptor disorders. Am J Hum Genet 2001;69:722–737.
40 Biel M, Seeliger M, Pfeifer A, Kohler K, Gerstner A, Ludwig A, Jaissle, G, Fauser S, Zrenner E, Hofmann F: Selective loss of cone function in mice lacking the cyclic nucleotide-gated channel CNG3. Proc Natl Acad Sci USA 1999;96:7553–7557.
41 Falls HF, Wolter JR, Alpern M: Typical total monochromacy. A histological and psychophysical study. Arch Ophthalmol 1965;74:610–616.
42 Glickstein M, Heath GG: Receptors in the monochromat eye. Vision Res 1975;15:633–636.

43 Hansen E. Clinical aspects of achromatopsia; in Hess RF, Sharpe LT, Nordby K (eds): Night Vision: Basic, Clinical and Applied Aspects. Cambridge, Cambridge University Press, 1990, p 337.
44 Morris TA, Fong SL: Characterization of the gene encoding human cone transducin α-subunit (GNAT2). Genomics 1993;17:442–448.
45 Cai K, Itoh Y, Khorana HG: Mapping of contact sites in complex formation between transducin and light-activated rhodopsin by covalent cross-linking: Use of a photoactivatable reagent. Proc Natl Acad Sci USA 2001;98:4877–4882.

Samir Deeb
Box 357720, University of Washington, Seattle, WA 98195 (USA)
Tel. +1 206 543 4001 Fax +1 206 543 0754, E-Mail deeb@genetics.washington.edu

Wissinger B, Kohl S, Langenbeck U (eds): Genetics in Ophthalmology.
Dev Ophthalmol. Basel, Karger, 2003, vol 37, pp 188–198

Gene Therapy and Animal Models for Retinal Disease

Nadine S. Dejneka, Tonia S. Rex, Jean Bennett

F.M. Kirby Center for Molecular Ophthalmology, Scheie Eye Institute,
University of Pennsylvania, Philadelphia, Pa., USA

Abstract

Those plagued by retinal diseases are often robbed of their vision, as often, effective treatments do not exist. Knowledge of the pathophysiology of retinal diseases stems from research on available animal models. Gene therapy may be useful for both genetic and acquired retinal diseases. This review will focus on retinal diseases for which gene therapy has demonstrated promise. The diseases are presented in order of the age at which they are generally first symptomatic and include retinopathy of prematurity, Leber congenital amaurosis, mucopolysaccharidoses, retinoblastoma, retinitis pigmentosa, diabetic retinopathy, glaucoma and age-related macular degeneration. We will describe the animal models used to study these disorders and emphasize the progress that has been made in using gene therapy for the treatment of retinal disease.

Introduction

The retina is a highly organized, complex, neural structure that is responsible for visual processing. Diseases affecting this tissue can be particularly devastating, as vision loss is irreversible and frequently severe. These diseases are often multifaceted, and treatments are limited by our incomplete understanding of the mechanisms that lead to blindness.

In order to further our understanding of acquired and inherited retinal diseases we depend heavily on animal models. Many models have arisen from spontaneous mutations, while others have been induced by environmental factors. Technological advances in genetic engineering have also enabled researchers to develop genetically engineered models that either overexpress genes known to lead to retinal disease or disrupt normal retinal gene function. These animals continue to expand our understanding of basic retinal biology and help to unravel the mysteries surrounding mechanisms of retinal disease.

Gene therapy may prove to be a feasible treatment option for patients with retinal diseases. Viral vectors can be used to transport therapeutic material to cells of the retina. Adenovirus (Ad), adeno-associated virus (AAV) and lentivirus are most commonly used for retinal transgene delivery. These vectors differ in their tropism for specific cell types; consequently, it is necessary to select a viral delivery system based on the ocular target [1]. Genes may themselves cause disease if they are targeted to the wrong cells. As different viruses vary in their onset and duration, these factors must be considered in selecting an appropriate delivery vehicle [1].

Here we review nine different disorders affecting the retina. They include retinopathy of prematurity, Leber congenital amaurosis, mucopolysaccharidosis, retinoblastoma, autosomal recessive retinitis pigmentosa, diabetic retinopathy, autosomal dominant retinitis pigmentosa, glaucoma and age-related neovascularization. These diseases are listed chronologically; according to the age symptoms first appear. We provide updates on the pathophysiology of each disease and provide an overview on results of gene therapy studies.

Retinopathy of Prematurity (ROP)

ROP is responsible for most cases of childhood blindness in the developed world, and specifically affects premature neonates [2]. Infants delivered at full term normally possess a completely vascularized retina. In contrast, premature neonates possess an incompletely vascularized retina. The incomplete vascularization coupled with the changes in oxygenation after birth often leads to a relative hypoxia in the avascularized retina. This stimulates the growth of new abnormal vessels at the juncture of the vascularized and avascularized retina. The vessels extend from the retina into the vitreous, causing hemorrhage and retinal detachment [2].

Current treatments for ROP are limited to cryosurgery and laser ablation of avascular retina [2]. The treatments are far from perfect as they themselves destroy functioning retina. Fortunately, a reproducible and quantifiable murine oxygen-induced retinopathy model of ROP exists, which may be used to test potential candidate treatments [3]. In this model, 1-week-old C57BL/6J mice are exposed to 75% oxygen for 5 days and then returned to room air. Retinal neovascularization results with a maximal response seen between postnatal days 17 and 21. Similar models have also been developed in rats [4, 5].

Antiangiogenic therapy may prove to be an effective treatment for ROP. Angiostatin, a proteolytic fragment of plasminogen, is a potent angiogenesis inhibitor and may contribute to the positive effects of photocoagulation therapy in neovascular retinal diseases [6]. An AAV vector carrying a shortened recombinant angiostatin derivative was used to produce angiostatin in vitro [7]. Following subsequent purification of the recombinant protein, subcutaneous delivery to the murine oxygen-induced retinopathy model resulted in effectively reducing blood vessel formation [7].

More recently, subretinal injection of an AAV carrying any of a variety of antiangiogenic genes including endostatin, pigment epithelium-derived factor (PEDF) and tissue inhibitor of metalloproteinases 3 (TIMP3) was found to significantly inhibit neovascular growth in the ROP model [8]. Further studies must be carried out in order to characterize the stability of the effect and to assure that the procedure is safe to the eye and to other organ systems.

Leber Congenital Amaurosis (LCA)

LCA is a childhood-onset retinal degeneration resulting in complete night blindness from birth or early childhood. Mutations in at least 6 genes have independently been shown to result in LCA [9]. One of these genes, RPE65 is, as its name suggests, localized to the retinal pigment epithelium (RPE), where recent evidence indicates it plays an essential role in the recycling of vitamin A [10].

Currently, two animal models of LCA exist. Both, as in a significant percentage of humans with LCA, have an RPE65 deficiency. The first model was identified as a spontaneous mutation in the Swedish Briard dog [11, 12]. Affected Briard dogs have a homozygous 4-bp deletion (485delAAGA) in putative exon 5 of the canine RPE65 gene, and suffer from the same severe visual impairments seen in human LCA [13, 14]. The second model, a knockout of RPE65, was developed in mice. These animals, like the Briards, suffer progressive retinal degeneration, have severely depressed visual function, lack the rhodopsin photopigment, and accumulate all-*trans* retinyl esters and droplets within the RPE [10, 15].

Recently, gene therapy has successfully recovered vision in the canine model of LCA [16]. The researchers designed an AAV carrying the wild-type canine *RPE65* cDNA and injected it into the subretinal space of dog eyes. They demonstrated functional recovery in the treated dogs by performing visual function and behavioral tests. Assuming that the therapeutic findings persist and there are no toxic consequences, a gene therapy trial may be developed to treat humans with the same genetic disease.

Mucopolysaccharidoses (MPSs)

MPSs are a group of inherited metabolic diseases characterized by the abnormal accumulation of glycosaminoglycans (heparan sulfate, dermatan sulfate, keratan sulfate) in the lysosomes of various tissues, as a result of defects in carbohydrate metabolism [17]. Patients have a short life expectancy and generally suffer disease in multiple organ systems. Ocular abnormalities range from corneal clouding to retinal degeneration, optic atrophy and glaucoma.

Multiple animal models of MPSs exist. These include a feline model of MPSVI (deficiency of arylsulfatase B), and murine, feline and canine models of MPSVII (deficiency of β-glucuronidase) [18, 19]. These models share multiple pathological abnormalities with humans including those seen in the eye.

To date, there have not been any successful treatment options for the ocular manifestations associated with MPS, although keratoplasty has had limited success in treating corneal opacification in an animal model of MPSIV [20]. In a promising gene therapy study, Li and Davidson [17] delivered a recombinant Ad carrying the human β-glucuronidase gene to the eyes of MPSVII mice. These mice develop late-onset photoreceptor degeneration secondary to defects in the RPE, and they possess lysosomal storage vacuoles in keratocytes in the corneal stroma, corneal endothelial cells, RPE cells, and cells in the choroid and sclera [17]. Intravitreal injection resulted in complete clearance of the storage defect in RPE cells, with partial phenotypic correction in corneal endothelial cells. Intravenous administration of a recombinant AAV encoding the human β-glucuronidase cDNA also resulted in nearly complete elimination of lysosomal storage vacuoles in the RPE of these animals [21]. In order to address the corneal clouding in the MPSVII mouse, Kamata et al. [22] delivered an Ad expressing human β-glucuronidase into the intrastromal region of the cornea following lamellar keratotomy. Histology revealed a rapid and nearly complete elimination of vacuoles in the areas of the cornea examined. Recently, a larger animal model has been used to explore the efficacy of treating MPSVI ocular disorders via gene therapy. Ho et al. [23] used an AAV to deliver arylsulfatase B to the subretinal space of the MPSVI cat. This model is characterized by the presence of vacuolated inclusions in the RPE, cornea, conjunctiva, sclera, choroid and the stroma of the iris and ciliary body [14]. AAV treatment appeared to reverse the diseased phenotype in the RPE [33].

Retinoblastoma

Retinoblastoma is an ocular tumor of childhood and is fatal if left untreated. Two thirds of all cases are diagnosed by 3 years of age, and tumors may present unilaterally or bilaterally. Tumor formation results from mutations in the retinoblastoma gene (*pRb*) [24]. This gene normally encodes a nuclear phosphoprotein that is important in regulating the cell cycle. In the heritable form of the disease, patients possess a single altered allele. When a spontaneous mutation disrupts the second retinoblastoma allele in retinal cells the tumor develops. Nonheritable disease occurs when both alleles are inactivated by spontaneous mutations.

Gene therapy studies have focused on developing an effective eye-sparing treatment for retinoblastoma. Current protocols use enucleation to treat large unilateral tumors in patients. Chemotherapy and radiation therapy may also be used, but are not always effective, and can cause further damage. An ideal therapy would potentially eliminate the need for such drastic treatments and preserve the integrity of the eye. Researchers have recently taken advantage of an Ad vector that encodes the herpes simplex virus thymidine kinase gene (*HStk*). Virus was directly administered to experimental tumors in mice, and animals were subsequently treated with ganciclovir [25]. Transduced cells were susceptible to ganciclovir cytotoxicity and tumor growth was inhibited. These results allowed for approval of a phase I clinical trial for the treatment of retinoblastoma. In this trial, ganciclovir is used to ablate the tumors that have been treated with AdV-thymidine kinase. This trial is currently underway, and early results appear promising [Hurwitz, pers. commun.].

Retinitis pigmentosa (Autosomal Recessive Disease)

Retinitis pigmentosa (RP) represents a group of inherited retinal disorders that affect approximately 1 in 3,000 individuals worldwide. RP is characterized by early-onset night blindness, which is followed by central vision loss. Patients with autosomal recessive (AR) disease often begin having symptoms in childhood or early adolescence. These symptoms become progressively worse with time, leading to complete visual impairment (age 30–60) [26].

AR RP can result as a consequence of lack-of-function mutations found in genes that play critical roles in the visual transduction cascade and outer segment maintenance [see 9]. An AR retinal degeneration occurs spontaneously in a number of animal models for RP, including the *rd* mouse, Irish setter dog, Royal College of Surgeons (RCS) rat, and the *rds* mouse. In order to rescue the lack of function phenotype it is necessary to introduce a correct version of the inactivated gene to its target cell. Gene therapy has served well for this purpose. βPDE was delivered to the *rd* mouse using Ad, AAV, gutted Ad and lentivirus [27–30]. These mice are homozygous for a mutation in βPDE and experience a rapid initial loss of rod photoreceptor cells that is followed by cone photoreceptor loss. Each of the four different virus treatments proved to successfully delay the rate of photoreceptor degeneration in the *rd* model. Similar approaches were used to treat the RCS rat. This animal has a mutation in the gene for the receptor tyrosine kinase Mertk [31]. These animals develop an abnormal build-up of outer segment debris in the subretinal space, due to the RPE's inability to phagocytize photoreceptor outer segments. This leads to a progressive loss of rod and cone photoreceptor cells [31]. Vollrath et al. [32] used subretinal injection to deliver an Ad encoding the rat *Mertk* cDNA to RCS rats. Treatment restored the RPE phagocytosis defect and preserved outer segment structure in areas surrounding the site of injection. The *rds* mouse possesses a lack of function mutation in the Prph2 (rds/peripherin) gene. This animal is a model for AR RP, although interestingly, *Prph2* mutations in humans with RP or macular dystrophy are inherited in an autosomal dominant (AD) fashion. Delivery of the wild-type *Prph2* gene to photoreceptors of the *rds* mouse has extended the lives and function of these cells [33].

One promising area of gene therapy for retinal degenerative diseases involves use of antiapoptotic genes, as photoreceptors ultimately die via the apoptotic cascade [34–36]. Ad-bcl-2 was subretinally delivered to *rd* mice [37]. Treatment resulted in a rescue effect that was maintained for 6 weeks. Such treatment may also be applicable to AD RP.

Diabetic Retinopathy

Diabetic retinopathy is the leading cause of blindness in individuals under the age of 65. Proliferative diabetic retinopathy (PDR) is characterized by the development of retinal capillary occlusions and small vessel damage. The capillaries develop microaneurysms and the retinal veins become tortuous and dilated, resulting in hemorrhage and retinal detachment causing sudden visual loss. The pathogenesis and symptoms of diabetic retinopathy appear very similar to those of ROP. Therefore, gene therapy strategies aimed at ROP are likely to apply to PDR, as well. Indeed, the ROP mouse is used as a model for studies involving treatment for both ROP and PDR.

Retinitis pigmentosa (Autosomal Dominant Disease)

AD RP may be caused by any of a large number of mutations in a number of photoreceptor-specific genes [9]. Multiple gene therapy approaches have been applied to rodent models of AD RP. Dominant disease results in unwanted gain of function, thus mutant protein must be eliminated in order for recovery to occur. Ribozymes are very useful for this purpose. These RNA enzymes are designed to exclusively target and cleave mutant mRNA. A ribozyme specific for the P23H mutant mRNA was delivered to the transgenic rhodopsin P23H rat via an AAV [38, 39]. Photoreceptor degeneration was successfully delayed for 3 months.

A second strategy for treating RP relies on the delivery of growth factors to promote photoreceptor survival or to limit apoptotic cell death. AAV delivery of basic fibroblast growth factor delayed photoreceptor death in the S344ter rhodopsin transgenic rat [40]. Delivery of ciliary neurotrophic factor resulted in long-term protection of retinal structures in the both rhodopsin S344ter and P23H animals [41]. AAV-mediated delivery of glial-derived neurotrophic factor also delayed cell death in S334ter transgenic rats [42].

Glaucoma

Glaucoma is the second leading cause of blindness in the world. It is an ocular dystrophy that is characterized by retinal ganglion cell degeneration, altered nitric oxide synthase levels, and elevated glutamate in the vitreous [for review, see 43]. The most common cause of glaucoma is impaired aqueous outflow from the anterior chamber, resulting in increased intraocular pressure (IOP). Interestingly, conventional treatment aimed at lowering IOP does not necessarily prevent ganglion cell death, suggesting other factors are involved.

A number of animal models for glaucoma have been developed in rodents. Two models that rely on mechanical interventions to increase IOP include episcleral vein cauterization and laser treatment of the trabeculum. Alternatively, optic nerve crush, optic nerve transection, or intravitreal injection of NMDA are performed as a means of inducing ganglion cell death.

Gene therapy for glaucoma has been limited to the delivery of growth factors and antiapoptotic proteins. Ad vectors were used to deliver brain-derived neurotrophic factor (BDNF) to a rat model of glaucoma [44]. Increased expression of BDNF in the Müller cells resulted in delayed retinal degeneration. AAVs encoding GFP were also used to target ganglion cells. The authors demonstrated that the ganglion cells are efficiently transduced and expression persists for up to 6 months [45]. Surprisingly, Dreyer et al. [46] illustrated that AAV-GFP alone was able to protect the ganglion cells from NMDA toxicity. However, intravitreal injection of AAV-bcl-2 in models of axonal injury and NMDA toxicity exacerbated retinal ganglion cell death [47], while AAV-bFGF appears to be protective [48].

Age-Related Macular Degeneration (AMD)

AMD is the leading cause of blindness in people over the age of 65. The 'wet' form of AMD results in the most damage – often leading to blindness overnight from hemorrhage under the retina originating from blood vessels which have aberrantly entered this area (after traversing Bruch's membrane and the RPE). Treatment is currently limited to laser photocoagulation of abnormal choroidal vessels. Unfortunately, this therapy is only suitable for a small subset of patients (~10%) and generally stabilizes vision loss for only a limited time.

It has been difficult developing an animal model of wet AMD. This is because, as yet, no genes have been conclusively identified as causing this disease and also because the only animal with a macula is the primate. Mechanical rupture of Bruch's membrane with laser photocoagulation will stimulate CNV in primates, rabbits and rodents; however, lesions generated in this way generally resolve spontaneously, unlike CNV found in humans with AMD. Attempts have also been made to generate transgenic rodents that over-express VEGF in photoreceptors. These animals develop neovascularization originating from the vitreal side of the retina – not the choroid [49, 50]. Despite the imperfect animal models for the wet form of AMD, they have been useful in testing potential treatments for this disorder. Mori et al. [51] recently delivered PEDF through Ad and found that this inhibits neovascularization in the transgenic VEGF mouse and the ROP model. This has led to the proposal of a human clinical trial testing this treatment in neovascular AMD.

Because of the general lack of animal models for wet AMD, gene therapy reagents have also been put to use to create an animal model so that medical treatments can be tested. Recently, an Ad vector encoding human VEGF was used to induce CNV in the Long-Evans rat [52]. This viral vector was delivered subretinally, and animals developed reproducible CNV two weeks post-injection, suggesting it may be a useful model to study neovascularization associated with AMD.

Conclusions

Retinal diseases are found at all stages of life and often cause significant morbidity. This review summarizes valuable animal models for a variety of retinal diseases and their roles in development of gene therapy-based treatments. Currently, rodents are the primary animals used to mimic human disease partly because of the abundance of spontaneous mutants that are available and also partly because of our ability to create genetically engineered mice. With ongoing advances in vector biology and the identification of novel genes it is possible that gene therapy will become an acceptable form of treatment for patients with at least certain forms of retinal disease.

References

1 Dejneka NS, Bennett J: Gene therapy and retinitis pigmentosa: Advances and future challenges. Bioessays 2001;23:662–668.
2 Neely KA, Gardner TW: Ocular neovascularization: Clarifying complex interactions. Am J Pathol 1998;153:665–670.
3 Smith LE, Wesolowski E, McLellan A, Kostyk SK, D'Amato R, Sullivan R, D'Amore PA: Oxygen-induced retinopathy in the mouse. Invest Ophthalmol Vis Sci 1994;35:101–111.
4 Penn JS, Henry MM, Tolman BL: Exposure to alternating hypoxia and hyperoxia causes severe proliferative retinopathy in the newborn rat. Pediatr Res 1994;36:724–731.
5 Cunningham S, McColm JR, Wade J, Sedowofia K, McIntosh N, Fleck B: A novel model of retinopathy of prematurity simulating preterm oxygen variability in the rat. Invest Ophthalmol Vis Sci 2000;41:4275–4280.
6 Spranger J, Hammes HP, Preissner KT, Schatz H, Pfeiffer AF: Release of the angiogenesis inhibitor angiostatin in patients with proliferative diabetic retinopathy: Association with retinal photocoagulation. Diabetologia 2000;43:1404–1407.
7 Meneses PI, Hajjar KA, Berns KI, Duvoisin RM: Recombinant angiostatin prevents retinal neovascularization in a murine proliferative retinopathy model. Gene Ther 2001;8:646–648.
8 Auricchio A, Behling KC, Maguire AM, O'Connor EE, Tolentino MJ, Wilson JM, Bennett J: Inhibition of retinal neovascularization by intraocular viral mediated delivery of anti-angiogenic agents. Mol Ther 2002;6:490–494.
9 RetNet, 2000. http://www.sph.uth.tmc.edu/retnet/
10 Redmond TM, Yu S, Lee E, Bok D, Hamasaki D, Chen N, Goletz P, Ma JX, Crouch RK, Pfeifer K: Rpe65 is necessary for production of 11-*cis*-vitamin A in the retinal visual cycle. Nat Genet 1998;20:344–351.
11 Narfstrom K, Wrigstad A, Nilsson SE: The Briard dog: A new animal model of congenital stationary night blindness. Br J Ophthalmol 1989;73:750–756.
12 Wrigstad A: Hereditary dystrophy of the retina and the retinal pigment epithelium in a strain of Briard dogs: A clinical, morphological and electrophysiological study. Linköping University Medical Dissertations 1994.
13 Veske A, Nilsson SE, Narfstrom K, Gal A: Retinal dystrophy of Swedish Briard/Briard-beagle dogs is due to a 4-bp deletion in RPE65. Genomics 1999;57:57–61.
14 Aguirre GD, Baldwin V, Pearce-Kelling S, Narfstrom K, Ray K, Acland GM: Congenital stationary night blindness in the dog: Common mutation in the RPE65 gene indicates founder effect. Mol Vis 1998;4:23.
15 Katz ML, Redmond TM: Effect of Rpe65 knockout on accumulation of lipofuscin fluorophores in the retinal pigment epithelium. Invest Ophthalmol Vis Sci 2001;42:3023–3030.
16 Acland GM, Aguirre GD, Ray J, Zhang Q, Aleman TS, Cideciyan AV, Pearce-Kelling SE, Anand V, Zeng Y, Maguire AM, Jacobson SG, Hauswirth WW, Bennett J: Gene therapy restores vision in a canine model of childhood blindness. Nat Genet 2001;28:92–95.
17 Li T, Davidson BL: Phenotype correction in retinal pigment epithelium in murine mucopolysaccharidosis VII by adenovirus-mediated gene transfer. Proc Natl Acad Sci USA 1995;92:7700–7704.
18 Haskins ME, Aguirre GD, Jezyk PF, Desnick RJ, Patterson DF: The pathology of the feline model of mucopolysaccharidosis I. Am J Pathol 1983;112:27–36.
19 Ray J, Wolfe JH, Aguirre GD, Haskins ME: Retroviral cDNA transfer to the RPE: stable expression and modification of metabolism. Invest Ophthalmol Vis Sci 1998;39:1658–1666.

20 Kasmann-Kellner B, Weindler J, Pfau B, Ruprecht KW: Ocular changes in mucopolysaccharidosis IV A (Morquio A syndrome) and long-term results of perforating keratoplasty. Ophthalmologica 1999;213:200–205.
21 Daly TM, Vogler C, Levy B, Haskins ME, Sands MS: Neonatal gene transfer leads to widespread correction of pathology in a murine model of lysosomal storage disease. Proc Natl Acad Sci USA 1999;96:2296–2300.
22 Kamata Y, Okuyama T, Kosuga M, O'Hira A, Kanaji A, Sasaki K, Yamada M, Azuma N: Adenovirus-mediated gene therapy for corneal clouding in mice with mucopolysaccharidosis type VII. Mol Ther 2001;4:307–312.
23 Ho T, Maguire A, Aguirre G, Salvetti A, Haskins M, Bennett J: Phenotypic rescue after adeno-associated virus-mediated delivery of arylsulfatase B to the retinal pigment epithelium of feline mucopolysaccharidosis VI. Invest Ophthalmol Vis Sci 2001;42:S348.
24 Cavenee WK, Dryja TP, Phillips RA, Benedict WF, Godbout R, Gallie BL, Murphree AL, Strong LC, White RL: Expression of recessive alleles by chromosomal mechanisms in retinoblastoma. Nature 1983;305:779–784.
25 Hurwitz MY, Marcus KT, Chevez-Barrios P, Louie K, Aguilar-Cordova E, Hurwitz RL: Suicide gene therapy for treatment of retinoblastoma in a murine model. Hum Gene Ther 1999; 10:441–448.
26 Phelan JK, Bok D: A brief review of retinitis pigmentosa and the identified retinitis pigmentosa genes. Mol Vis 2000;6:116–124.
27 Bennett J, Tanabe T, Sun D, Zeng Y, Kjeldbye H, Gouras P, Maguire AM: Photoreceptor cell rescue in retinal degeneration (rd) mice by in vivo gene therapy. Nat Med 1996;2:649–654.
28 Jomary C, Vincent KA, Grist J, Neal MJ, Jones SE: Rescue of photoreceptor function by AAV-mediated gene transfer in a mouse model of inherited retinal degeneration. Gene Ther 1997;4: 683–690.
29 Kumar-Singh R, Farber DB: Encapsidated adenovirus mini-chromosome-mediated delivery of genes to the retina: Application to the rescue of photoreceptor degeneration. Hum Mol Genet 1998;7:1893–1900.
30 Takahashi M, Miyoshi H, Verma IM, Gage FH: Rescue from photoreceptor degeneration in the rd mouse by human immunodeficiency virus vector-mediated gene transfer. J Virol 1999;73: 7812–7816.
31 D'Cruz PM, Yasumura D, Weir J, Matthes MT, Abderrahim H, LaVail MM, Vollrath D: Mutation of the receptor tyrosine kinase gene Mertk in the retinal dystrophic RCS rat. Hum Mol Genet 2000;9:645–651.
32 Vollrath D, Feng W, Duncan JL, Yasumura D, D'Cruz PM, Chappelow A, Matthes MT, Kay MA, LaVail MM: Correction of the retinal dystrophy phenotype of the RCS rat by viral gene transfer of Mertk. Proc Natl Acad Sci USA 2001;98:12584–12589.
33 Ali RR, Sarra GM, Stephens C, Alwis MD, Bainbridge JW, Munro PM, Fauser S, Reichel MB, Kinnon C, Hunt DM, Bhattacharya SS, Thrasher AJ: Restoration of photoreceptor ultrastructure and function in retinal degeneration slow mice by gene therapy. Nat Genet 2000;25:306–310.
34 Chang GQ, Hao Y, Wong F: Apoptosis: Final common pathway of photoreceptor death in rd, rds, and rhodopsin mutant mice. Neuron 1993;11:595–605.
35 Portera-Cailliau C, Sung CH, Nathans J, Adler R: Apoptotic photoreceptor cell death in mouse models of retinitis pigmentosa. Proc Natl Acad Sci USA 1994;91:974–978.
36 Lolley RN, Rong H, Craft CM: Linkage of photoreceptor degeneration by apoptosis with inherited defect in phototransduction. Invest Ophthalmol Vis Sci 1994;35:358–362.
37 Bennett J, Zeng Y, Bajwa R, Klatt L, Li Y, Maguire AM: Adenovirus-mediated delivery of rhodopsin-promoted bcl-2 results in a delay in photoreceptor cell death in the rd/rd mouse. Gene Ther 1998;5:1156–1164.
38 Lewin AS, Drenser KA, Hauswirth WW, Nishikawa S, Yasumura D, Flannery JG, LaVail MM: Ribozyme rescue of photoreceptor cells in a transgenic rat model of autosomal dominant retinitis pigmentosa. Nat Med 1998;4:967–971.
39 LaVail MM, Yasumura D, Matthes MT, Drenser KA, Flannery JG, Lewin AS, Hauswirth WW: Ribozyme rescue of photoreceptor cells in P23H transgenic rats: Long-term survival and late-stage therapy. Proc Natl Acad Sci USA 2000;97:11488–11493.

40 Lau D, McGee LH, Zhou S, Rendahl KG, Manning WC, Escobedo JA, Flannery JG: Retinal degeneration is slowed in transgenic rats by AAV-mediated delivery of FGF-2. Invest Ophthalmol Vis Sci 2000;41:3622–3633.

41 Liang FQ, Aleman TS, Dejneka NS, Dudus L, Fisher KJ, Maguire AM, Jacobson SG, Bennett J: Long-term protection of retinal structure but not function using RAAV.CNTF in animal models of retinitis pigmentosa. Mol Ther 2001;4:461–472.

42 McGee Sanftner LH, Abel H, Hauswirth WW, Flannery JG: Glial cell line derived neurotrophic factor delays photoreceptor degeneration in a transgenic rat model of retinitis pigmentosa. Mol Ther 2001;4:622–629.

43 Farkas RH, Grosskreutz CL: Apoptosis, neuroprotection and retinal ganglion cell death: An overview. Int Ophthalmol Clin 2001;41:111–130.

44 Di Polo A, Aigner LJ, Dunn RJ, Bray GM, Aguayo AJ: Prolonged delivery of brain-derived neurotrophic factor by adenovirus-infected Müller cells temporarily rescues injured retinal ganglion cells. Proc Natl Acad Sci USA 1998;95:3978–3983.

45 Dudus L, Anand V, Acland GM, Chen SJ, Wilson JM, Fisher KJ, Maguire AM, Bennett J: Persistent transgene product in retina, optic nerve and brain after intraocular injection of rAAV. Vision Res 1999;39:2545–2553.

46 Dreyer EB, Vorwerk CK, Zurakowski D, Simon PD, Bennett J: Infection with adeno-associated virus may protect against excitotoxicity. Neuroreport 1999;10:2887–2890.

47 Simon PD, Vorwerk CK, Mansukani SS, Chen SJ, Wilson JM, Zurakowski D, Bennett J, Dreyer EB: bcl-2 gene therapy exacerbates excitotoxicity. Hum Gene Ther 1999;10:1715–1720.

48 Vorwerk CK, Dejneka NS, Schuettauf F, Bennett J, Naskar R, Dreyer EB: In vivo gene therapy with AAV-bFGF protects against excitotoxicity. Invest Ophthalmol Vis Sci 2000;41:S16.

49 Okamoto N, Tobe T, Hackett SF, Ozaki H, Vinores MA, LaRochelle W, Zack DJ, Campochiaro PA: Transgenic mice with increased expression of vascular endothelial growth factor in the retina: A new model of intraretinal and subretinal neovascularization. Am J Pathol 1997;151:281–291.

50 Schwesinger C, Yee C, Rohan RM, Joussen AM, Fernandez A, Meyer TN, Poulaki V, Ma JJ, Redmond TM, Liu S, Adamis AP, D'Amato RJ: Intrachoroidal neovascularization in transgenic mice overexpressing vascular endothelial growth factor in the retinal pigment epithelium. Am J Pathol 2001;158:1161–1172.

51 Mori K, Duh E, Gehlbach P, Ando A, Takahashi K, Pearlman J, Yang HS, Zack DJ, Ettyreddy D, Brough DE, Wei LL, Campochiaro PA: Pigment epithelium-derived factor inhibits retinal and choroidal neovascularization. J Cell Physiol 2001;188:253–263.

52 Baffi J, Byrnes G, Chan CC, Csaky KG: Choroidal neovascularization in the rat induced by adenovirus-mediated expression of vascular endothelial growth factor. Invest Ophthalmol Vis Sci 2000;41:3582–3589.

Jean Bennett
Scheie Eye Institute, University of Pennsylvania,
310 Stellar-Chance Labs,
422 Curie Blvd, Philadelphia, PA 19104-6069 (USA)
Tel. +1 215 8980915, Fax +1 215 5737155, E-Mail jebennet@mail.med.upenn.edu

Wissinger B, Kohl S, Langenbeck U (eds): Genetics in Ophthalmology.
Dev Ophthalmol. Basel, Karger, 2003, vol 37, pp 199–214

Support for Patients Loosing Sight

Susanne Trauzettel-Klosinski, Gesa-Astrid Hahn

University Eye Hospital, Tübingen, Germany

Abstract

This overview on support of patients loosing sight is based on a literature survey regarding reading disabilities and orientation and on results of experience trials performed at the University Eye Clinic Tübingen. In reading disorders, the main goal of rehabilitation is to regain or maintain the ability to read newspaper print. The fundament of rehabilitation is the use of optical and electronical devices and the application of specially designed training programs. The ability of a person with low vision to achieve successful orientation and mobility rehabilitation depends on residual vision, posture and balance, body image, auditory and tactile abilities, intelligence and personality. Rehabilitation efforts focus on the enhancement of residual vision applying magnifying contrast-enhancing and photomultiplying devices. The main pillar of orientation and mobility rehabilitation is a training especially designed for the patient's needs. Rehabilitation efforts must be tailored to the type of vision loss and to specific functional implications – the success rate is high. An optimal fitting of the required spectrum of low vision aids should be provided to the patient and importantly, professional teaching and training is recommended. Activities of daily living, orientation and mobility, and psychological concerns must be addressed. Close cooperation with other branches of rehabilitation is essential.

Introduction: Aspects of Vision Loss

Relevant visual impairment is defined as a visual acuity (VA) of ≤0.3. Legal blindness is defined as ≤0.1 VA in some countries (e.g., USA, Sweden), as ≤0.02 in others (e.g., Germany) – or severe visual field (VF) defects (5° radius), even with normal VA.

Independently of these differences of definition, which are related to different monetary benefits, it is obvious that a legally blind person can have some residual vision that can be extremely valuable. In the case of a permanent deficit,

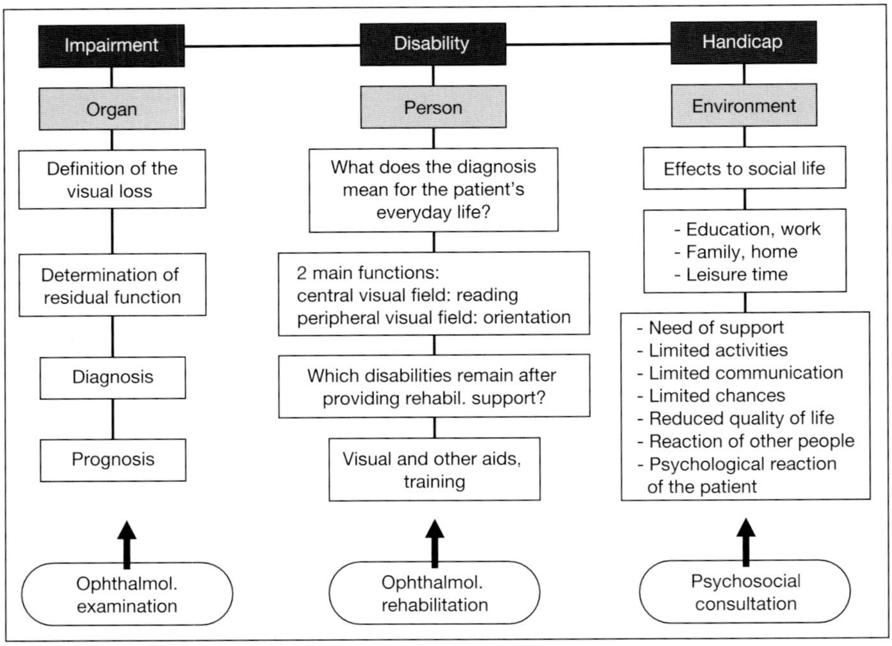

Fig. 1. Three aspects of vision loss based on the International Classification of Impairments, Disabilities and Handicaps [1].

three aspects have to be considered – based on the International Classification of Disabilities by the World Health Organization [1, 2]: (1) *Impairment:* damage of the organ; (2) *Disability:* functional restriction of the person, and (3) *Handicap:* social restriction related to the environment. These three aspects are related to the visual system shown in figure 1.

The impairment of the organ (eye, visual pathways) is described by the definition of the kind and extent of the visual loss (VA, VF, contrast sensitivity, etc.). It is important to determine the residual function, to state an unambiguous diagnosis and to estimate the prognosis.

After providing this information by the ophthalmological examination, the significant question is: Which disabilities result from the impairment for the individual? What are the consequences for everyday life?

In principle, there are two main functions in everyday life: The most important function of the central VF is reading, that of the peripheral VF is orientation.

Rehabilitation programs should specifically address these disabilities. After providing visual or other aids or training, it is important to check what degree of compensation was achieved and, conversely, what disabilities remain after rehabilitation.

Disabilities cause handicaps in the social environment: The patient's social life can be affected regarding education, work, family and activities during leisure time. Handicapped persons need support in various ways due to the limitations imposed on their activities and their communication. Thus, the patients can end up with fewer chances and a reduced quality of life. Their well-being also depends on the reactions of other people to the handicap and the patient's own psychological reaction. A psychosocial consultation with the aim to improve coping with all these problems is very important.

Reading Disorders

Reading is a crucial ability in modern society and means independence, communication, mental agility and quality of life.

Physiological Preconditions of Reading

With increasing eccentricity, VA decreases rapidly [3]. To read newspaper print at a distance of 25 cm, a VA of about 0.4 is required (fig. 2a). However, testing VA alone is not sufficient to measure reading ability, because it assesses the recognition of only one optotype at a time. During one fixation, the simultaneous perception of a whole group of letters is necessary, which requires a functioning VF of a minimum extent. This 'minimum VF for reading' extends 2° each to the right and left of fixation and 1° above and below [4] and corresponds more or less to the 'word recognition span' [5] and to the 'visual span' [6].

When these data are related to a text (fig. 2b), it is obvious that the text is perceived clearly only within this minimum reading VF. However, the 'total perceptual span' during one fixation can be widened up to 5° in reading direction. This provides a preview benefit by using parafoveal information processing [7].

To see the next letter group clearly, the person has to make a saccade. Reading means a regular pattern of saccades and fixations along the line and a return sweep to the beginning of the next line. This results in a typical stairstep pattern in an eye movement recording.

Reading in Pathological Conditions

Diseases of the eyes and visual pathways, which are associated with lesions of the central retina or defects in the central VF, respectively, cause reading disorders [8–11; for an overview, see 12] (table 1).

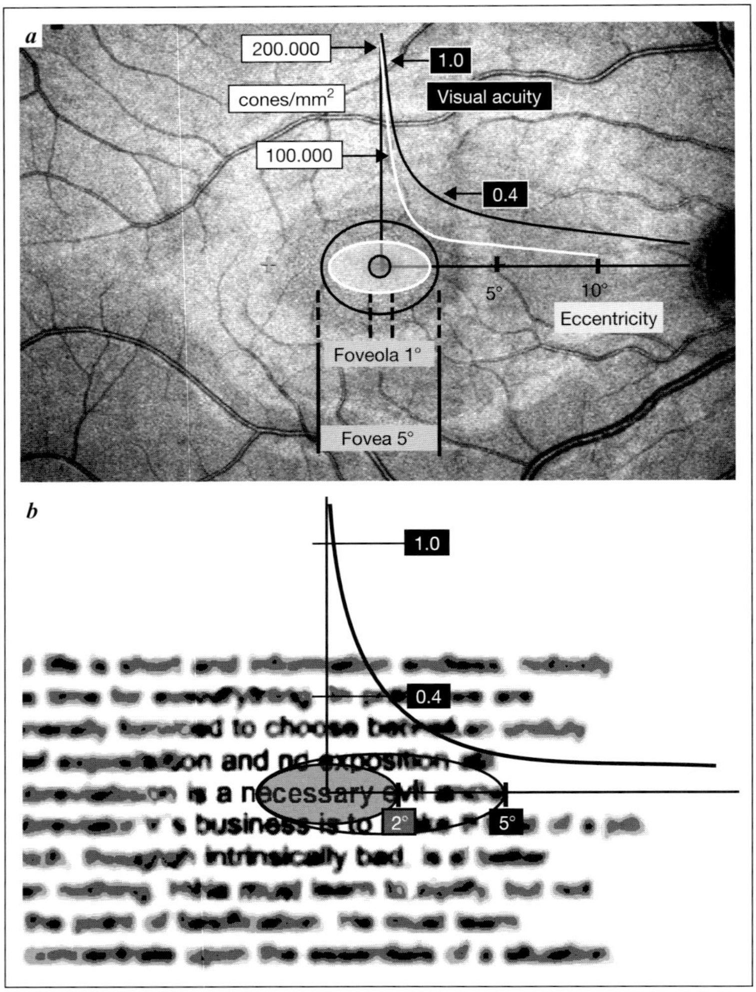

Fig. 2. ***a*** Morphological and functional data displayed on an SLO-fundus image: VA (dark line) and cone density (bright line) dependent on eccentricity, the proportions of the foveola and fovea, the minimum VF for reading (inner ellipse, 2° to the right and left and 1° above and below fixation) (modified after Trauzettel-Klosinski et al. [14]) ***b*** VA, minimum reading VF (inner ellipse) and total perceptual span (outer ellipse) related to the text. For reading newspaper print in 25 cm, a VA of 0.4 and a minimum reading VF of 2° to each side of fixation is necessary. The total perceptual span during one fixation can be extended up to 5° in reading direction (modified after Trauzettel-Klosinski [12]).

Table 1. Effect of the main hereditary eye diseases to reading ability

Diagnosis	Main feature	Foveal function	Fixation	Problem	Solution
Cataract Corneal dystrophy	Opaque media	Intact	Central	Contrast (resolution)	Contrast enhancement (plus magnification)
Maculopathy Optic atrophy	Central scotoma	Lost	Eccentric	Resolution	Magnification
Maculopathy Retinal dystrophy	Ring scotoma	Intact	Central	Size of visual field for reading	Wait for or train eccentric fixation
Primary peripheral retinal dystrophy Glaucoma (advanced)	Concentric field	Intact or reduced	Central	Resolution and size of visual field for reading	Contrast enhancement

(a) Patients with opaque optic media (cataract, corneal dystrophy) are mainly disabled by reduced contrast (fig. 3.1b). Lowered VA and glare are additional symptoms. Contrast enhancement plus magnification are necessary to restitute reading ability.

(b) Diseases of the central retina, especially maculopathies, cone dystrophy, achromatopsia, and most optic atrophies, especially Leber's and autosomal dominant OA, cause a central scotoma. This covers the minimum reading VF and makes it useless (fig. 3.2a). These patients can learn a valuable adaptive strategy: eccentric fixation or eccentric viewing (fig. 3.2b), where they use a healthy retinal area near the edge of the scotoma. This preferred retinal locus (PRL) is now the new center of the VF. Therefore, the scotoma and blind spot, which serves as a reference, are shifted [13, 14]. The resolution of the PRL is insufficient to read newspaper print. However, if the text is magnified, reading ability is regained (fig. 3.2c). This is the basis for the application of magnifying visual aids. A ring scotoma occurs in diseases of the central retina, which spares a small seeing central island within a central scotoma, especially in intermediate stages of Stargardt's disease (fig. 3.3a). The central island can become too small for reading, so that magnification does not help, except if the patient learns to use an eccentric PRL, as seen in figure 3.2b, despite intact foveal function, which is very difficult. The presence of a ring scotoma explains a discrepancy between good VA (single optotype recognition) and reading disability (insufficient size of the reading VF) [15].

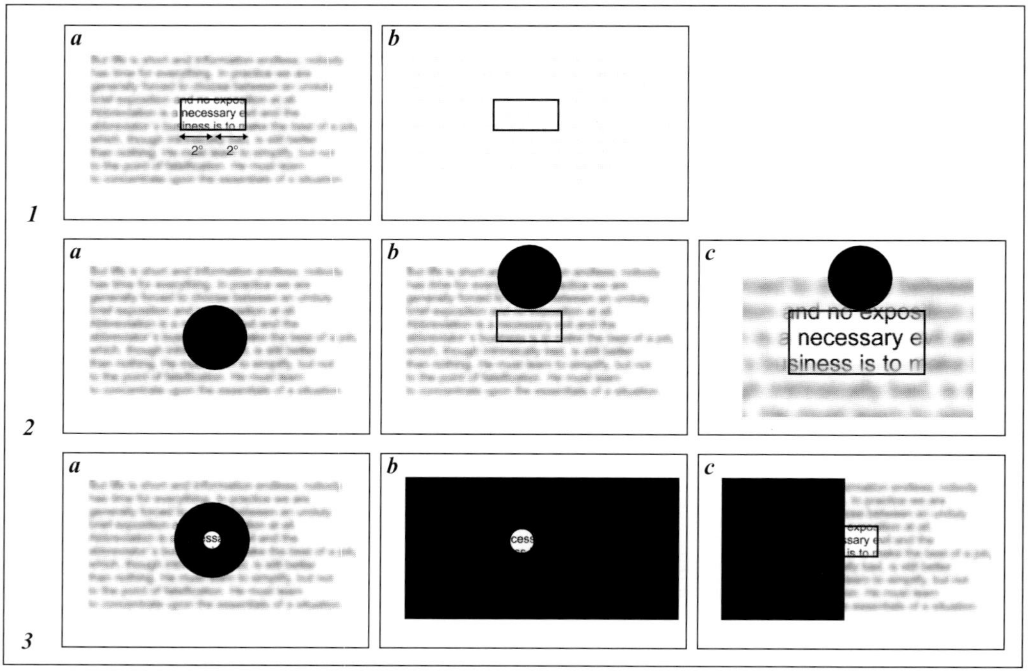

Fig. 3. The effect of different eye diseases on the reading process. ***1a*** Normal situation: within the minimum reading VF the text is perceived clearly. ***1b*** Opaque optic media: reduction of contrast causes the text to appear blurred. ***2a–c*** Central scotoma: ***a*** In central fixation, the reading VF is covered by the scotoma and useless. ***b*** In eccentric fixation, a healthy retinal area at the margin of the scotoma is used. This new PRL is now the center of the VF. Therefore, the scotoma is shifted. The resolution of the PRL is insufficient for reading newspapers print. ***c*** When the text is magnified, reading ability is regained. ***3a*** Ring scotoma: the central island of vision within a central scotoma can become too small for reading. Only if the patient learns to use an eccentric PRL for reading (as in ***2b***) will he or she be able to read magnified text. This explains a discrepancy between good visual acuity (single optotype) and reading inability (insufficient size of reading VF). ***3b*** In concentric VF the central island can become too small for reading. ***3c*** In hemianopia half of the reading VF is covered.

(c) Diseases of the peripheral retina, especially retinitis pigmentosa and rod-cone dystrophy, and advanced glaucoma lead to a concentrically restricted field, which can become too small for reading (fig. 3.3b). If VA is high enough, text size reduction and contrast enhancement can restitute reading ability.

(d) Visual pathway defects due to post-chiasmatic lesions cause homonymous hemianopia, which also limits the size of the reading VF (fig. 3.3c). The crucial condition is the distance of the field defect from the vertical midline, especially the presence of macular sparing. Some of these patients are also able to use an eccentric PRL in order to shift their field defect towards the hemianopic side and to create a useful reading VF [16, 17].

Rehabilitation in Patients with Reading Disorders

The aim of rehabilitation by optimization of the residual function is maintenance, restitution or optimization of the ability to read newspaper print.

The main tools are visual aids and training.

Visual Aids

Magnification. There is a wide spectrum of magnifying visual aids available, e.g. hand-held magnifiers, magnifying spectacles and CCTV devices. The choice of a particular aid depends on magnification need, task, working distance, field of view and the patient's own capabilities and motivation.

Contrast Enhancement. Optimal illumination of the reading material and electronic devices, such as CCTV monitors, can improve the contrast, which is indicated in patients who cannot be helped by magnification (insufficient size of reading VF), especially in retinitis pigmentosa. The contrast-enhancing cut-off filters are rarely used for reading.

Training

Training can take a variety of approaches: (a) training to handle the visual aid is crucial for its optimal use; (b) reading training: to stay in good practice as a general training; (c) specific reading programs to improve the reading process will be of increasing significance in future rehabilitation programs [18], and (d) training to use a favorable PRL can be helpful [19].

Other Aids

If reading cannot be acquired visually, auditory (synthesized voice, audio recordings) or tactile (Braille) aids can be provided. For education and work, complete electronic workstations can be extremely valuable for visually impaired persons.

Aids for Distance Reading

Patients benefit from hand-held telescopes as a mobile aid to read signs at bus stops and street signs, schoolchildren can use them to read from the blackboard. At school, a video-camera connected to the CCTV monitor can also be very helpful.

Success Rate of Low Vision Aids

An attempt at rehabilitation is always worthwhile – the success rate is high. In a cohort of 763 low vision patients of the Low Vision Clinic Tübingen, reading ability was present in only 13% of the patients before consultation, but in 90% after consultation. The success was not dependent on age, but on the functional status and on the motivation of the patient [20].

Orientation Disability

Physiological Preconditions of Orientation

The visual, vestibular and somatosensory systems collectively yield spatial orientation providing data for goal-directed action (fig. 4). Disorders that have an impact on the integrity of a single part of this neural network can lead to deficits of orientation and mobility and, thus, to disability. Simple activities of daily living become seriously challenged.

Orientation in Pathological Conditions

(a) Patients suffering from maculopathies (such as Stargardt's or Best's disease) or optic atrophies (as Leber's optic atrophy or autosomal dominant optic atrophy) are handicapped through diminished VA due to a central scotoma. This reduces the ability to read, while residual navigational vision remains.

(b) Diseases affecting the integrity of the peripheral VF have a serious impact on orientation and mobility: In the early stages of retinitis pigmentosa night blindness, diminished contrast sensitivity and loss of peripheral vision are characteristic symptoms. With disease progression, the patient's VF diminishes to a tubular remnant with a temporal island. Choroideremia is a rare progressive dystrophy of the choroid leading to severe VF loss. In the final stages, both diseases lead to blindness.

(c) Patients suffering from congenital achromatopsia are severely handicapped. In this disorder the cone photoreceptors are affected. Patients are almost or totally color blind and have poor VA (patients use their rod photoreceptors, which saturate at higher levels of illumination), the extent of the visual deficit varies.

(d) Congenital stationary night blindness disables the patient to navigate in the dark. The disease is characterized as a stationary retinal dysfunction. Leading to electrophysiological changes affecting mainly the rod function.

(e) Some forms of hereditary optic atrophy present with combined disabilities affecting neurological functions beyond the visual system: The infantile hereditary optic atrophy is associated with deficits such as ataxia or auditory loss. In optic atrophy of the 'Behr' type the patients are mentally retarded and

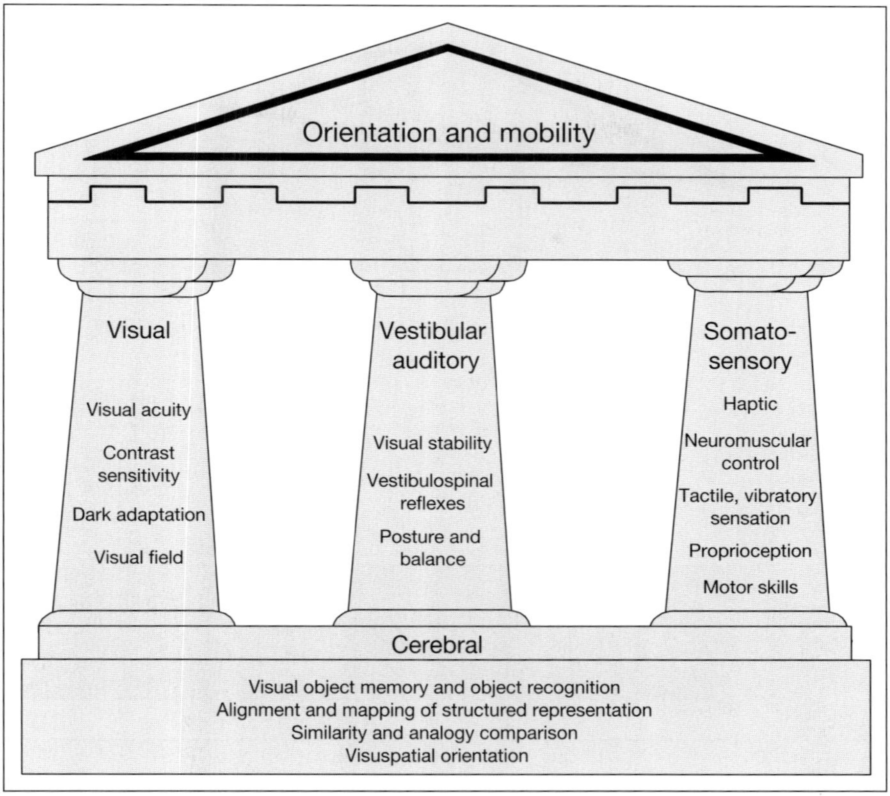

Fig. 4. Orientation and mobility are based on psychophysical functions including visual, vestibular, auditory and sensorimotor perception, cortical parameters of memory and learning that lead to goal directed action, usually in form of movement.

suffer from motor dysregulation. Also, neurodegenerative disorders such as Friedreich's ataxia with progressive cerebellar degeneration and deafness present with optic atrophy and RPE degeneration. Patients with gangliosidosis type 1 and 2 suffer progressive neurological deficit and early blindness within the first year of life due to optic atrophy. Neuronal ceroid lipofuscinosis is a further example of a storage disease leading to a progressive neurodegenerative disorder with seizures, optic atrophy, maculopathy and RPE degeneration.

Rehabilitation in Patients with Orientation Disability

The main therapeutical approach is to strengthen the remaining pillars of orientation and mobility by initiating a training plan for orientation and mobility

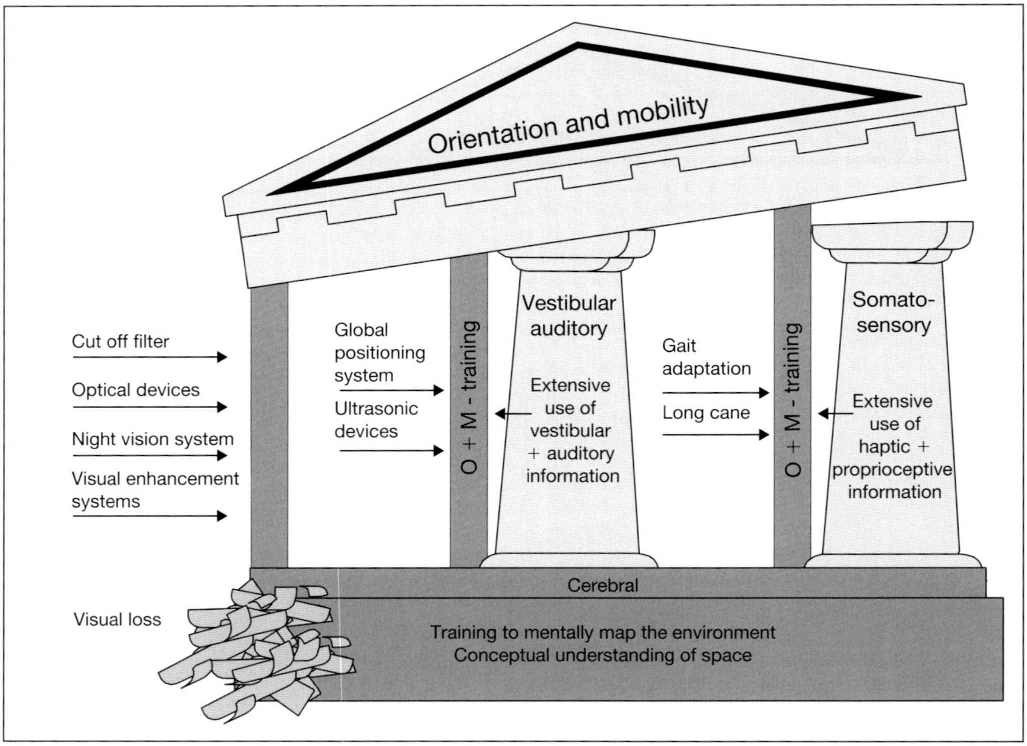

Fig. 5. Behavioral neural plasticity is important in the process of adapting visual alteration. Training and the use of remaining intact systems are used to profit from and to enhance neuronal plasticity.

(fig. 5). As outlined above, the complete battery of neurophysiological functions might not be at hand for rehabilitation of visually impaired patients.

Low Vision Aids

Cut-Off Filters

Cut-off filters block a defined part of the wavelengths of the light spectrum. This results in the characteristic transmission graph shown in figure 6 with an edge at defined wavelengths. Short-pass and long-pass edge filters are defined according to the cut-off wavelength.

Light perception of receptors in the zone of spectral transmission is not reduced, whereas receptors with differing spectrum within the cut-off effect are affected. This difference of light exposure and receptor activation is responsible for the effect of contrast enhancement reported by the patients [21]. Wearing

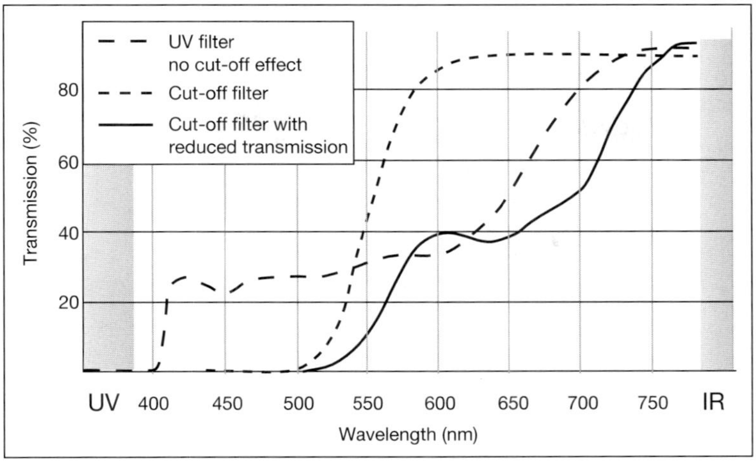

Fig. 6. Cut-off filters block a defined part of the wavelengths of the light spectrum whereas regular filters reduce the transmission of all wavelengths. This results in the characteristic transmission graph with an edge at defined wavelengths. Filters with short-pass and long-pass edge are defined according to the cut-off wavelength.

cut-off filters at high luminance levels eases the adaptation process in the change to lower luminance levels once the filter glasses have been taken off. Intact retinal layers and receptors compensate for the filter effects [22]. This might explain the more pronounced effect of cut-off filters in diseases of the outer and middle retinal layers [23]. Cut-off filters are recommended in retinitis pigmentosa and other dystrophic retinal diseases. Patients with aniridia or albinism gain functional improvement [23].

Application and evaluation protocol: Studies have assessed effects of cut-off filters on VA, contrast sensitivity, VF, adaptation time, glare and photophobia. Very little objective evidence has been provided to support anecdotal reports of improvements in visual performance [24].

(a) The evaluation of contrast sensitivity at various levels of luminance helps to identify those patients who will benefit from cut-off filters [25]. In patients with retinitis pigmentosa, Rüther et al. [26] found that patients with low contrast sensitivity showed relatively high benefit from the filters, especially with low spatial frequencies. Similar effects were found by Wetzel et al. [27].

(b) Perimetry can be used to evaluate improvement due to the filter quantitatively [28].

The residual and intact cone type determines the filter type and the long- or short-pass edge characteristics. Yet color testing as only parameter is not

a b

Fig. 7. a, b An empirical, individual testing of cut-off filters is recommended. Impression of 3 children with achromatopsia previous to and after fitting of cut-off filters.

sufficient to determine the filter type to be used. Also, the fitting of the filter has to account for a possibly unfavorable effect on the color vision which is anyhow diminished due to the underlying disease [21]. An empirical and individual fitting of the various types of cut-off filters is recommended (fig. 7).

Night Vision Aids

During daylight and in the first phase of dark adaptation, the cone system performs color discrimination and high spatial resolution. In the dark-adapted eye the monochromatic rod visual system is much more sensitive. Consequently, loss of rod function results in serious disability to navigate and move in the dark. Night vision devices are comparable to those developed for military purposes using photomultipliers to enhance the ambient light – these aids utilize the functionally more intact cone system.

The first studies were performed with hand-held devices [29, 30]. Friedburg et al. [31] evaluated the practical use of NIVIS, one example of night vision spectacles. Patients showed better VA, contrast sensitivity and motion perception at low and high scotopic levels using the night vision aid. The authors reported a short imaging delay between camera and display which might result in balance problems in patients with impairment of the vestibular system. Under glare conditions, the applicability of the device was limited. Rohrschneider et al. [32] evaluated the possibilities of rehabilitation with a further developed night vision device (Davis) performing typical outdoor activities. They found enhancement of contrast acuity during night and twilight conditions. Patients reported an improvement of orientation leading to an increase of the walking speed.

Practical application: (a) Patients with impaired rod visual system can profit using a night vision device. But an additional impairment of the cone visual system will limit the possible improvement using this system. Friedburg et al. [31] and Rohrschneider et al. [32] found that a minimum VA of 20/200 (0.1) was needed for a successful use of the device. On the other hand, patients with good VA might find the resolution of the display inferior to their actual VA [31]. (b) For successful use, the residual tubular VF should have at least 6–10° radius. Since the image projected on the display represents only a small detail of a normal VF, patients with night blindness and only minimal limitation of the VF might find this system not helpful. (c) Individual testing and adjustment of the device is necessary, and, above all, individual training of orientation and mobility is inevitable for successful use.

Low Vision Enhancement System

This device was designed to gain optimal utilization of residual vision by enlarging and contrast-enhancing images of the environment and print. It allows focussing at different working distances combined with variable magnification [33, 34]. This device was found to be useful for patients suffering from a central scotoma accompanied by a magnification need <8 times. Patients with a need for adaptation to different working distances exceeding the range of optical aids might profit in using this aid. With its specific clinical use, this type of visual aid is not in widespread use.

Extensive Use of Vestibular and Auditory Information

The extensive use of vestibular and auditory information is a major part of the training for orientation and mobility [35].

Haptic and Proprioceptive Learning

Haptic and proprioceptive learning is used to enhance non-visual motor skills: The training of gait adaptations is necessary to adapt to environmental changes and enables subjects to maintain safe mobility. Exploration of the environment is further extended using technical aids and a long cane to gain tactile information (obstacles, dimension of objects) [36]. Electronic mobility aids combined with ultrasonic wave processing convert inaudible ultrasound echoes into audible sound via the Doppler effect, to provide the user with information regarding his movement and spatial location [37]. A future option might be a system of beacons and personal receivers. The satellite-based global positioning system (GPS) is already in widespread use. This concept has been used in the 'Easy Walker' system available in The Netherlands. Kooijman and

Uyar [38] evaluated if these features are helpful for the mobility of visually impaired and found that much independence could be gained if such a system was available for visually impaired patients.

Neuronal Plasticity

With a condition of severely reduced visual information, training is focussed on object recognition and carrying out similarity and analogy comparisons without primarily using visual information. Patients are instructed to mentally map the environment and to gain a conceptual understanding of space [39].

Concluding Remarks

An attempt at rehabilitation is always worthwhile. Low vision rehabilitation has dramatic results in increasing the independence of affected persons. Rehabilitation efforts must be tailored to the type of vision loss and to specific functional implications. Special emphasis lies on training and practice with low vision aids and with orientation and mobility. Further options will arise from the increasing application of computer technology.

References

Reading

1 World Health Organization: Classification of Impairments, Disabilities and Handicaps. Geneva, WHO, 1980.

2 Colenbrander, G: Visual Standards – Aspects and Ranges of Vision Loss. Report to the International Council of Ophthalmology, Sydney, April 2002.

3 Wertheim T: Über die indirekte Sehschärfe. Z Psychol 1894;7:172–187.

4 Aulhorn E: Über Fixationsbreite und Fixationsfrequenz beim Lesen gerichteter Konturen. Pflügers Arch Physiol 1953;257:318–328.

5 Rayner K, Well AD, Pollatsek A: The availability of useful information to the right of fixation in reading. Percept Psychophys 1982;31:537–544.

6 Legge GE, Ahn SJ, Klitz TS, Luebker A: Psychophysics of reading. XVI. The visual span in normal and low vision. Vision Res 1997;37:1999–2010.

7 McConkie GW, Rayner K: The span of the effective stimulus during a fixation in reading. Percept Psychophys 1975;17:578–586.

8 Cummings RW, Whittaker SG, Watson GR: Scanning characters and reading with a central scotoma. Am J Optom Physiol Opt 1985;62:833–843.

9 Fletcher DC, Schuchard RA, Watson G: Relative locations of macular scotomas near the PRL: Effect on low vision reading. J Rehabil Res Dev 1999;36:356–364.

10 Timberlake GT, Mainster MA, Peli E et al: Reading with a macular scotoma. I. Retinal location of scotoma and fixation area. Invest Ophthalmol Vis Sci 1986;27:1137–1147.

11 Timberlake GT, Peli E, Essock EA, Augliere RA: Reading with a macular scotoma. II. Retinal locus for scanning text. Invest Ophthalmol Vis Sci 1987;28:1268–1274.

12 Trauzettel-Klosinski S: Reading disorders due to visual field defects – A neuro-ophthalmological view. Neuroophthalmology 2002;52:in press.

13 Aulhorn E: Die Gesichtsfeldprüfung bei makularen Erkrankungen; in Bericht der Zusammenkunft der DOG, Heidelberg 1973. München, Bergmann, 1975, pp 77–86.
14 Trauzettel-Klosinski S, Teschner C, Tornow RP, Zrenner E: Reading strategies in normal subjects and in patients with macular scotoma – assessed by two new methods of registration. Neuroophthalmology 1994;14:15–30.
15 Trauzettel-Klosinski S, Tornow RP: Fixation behaviour and reading in patients with macular scotoma – assessed by Tübingen Manual Perimetry and Scanning Laser Ophthalmoscopy. Neuroophthalmology 1996;16:241–253.
16 Trauzettel-Klosinski S: Eccentric fixation with hemianopic field defects. A valuable strategy to improve reading ability and an indication of cortical plasticity. Neuroophthalmology 1997;18:117–131.
17 Trauzettel-Klosinski S, Reinhard J: The vertical field border in hemianopia and its significance for fixation and reading. Invest Ophthalmol Vis Sci 1998;39:2177–2186.
18 Hahn GA, Stockum A, Teschner C, Mackeben M, Trauzettel-Klosinski S: Reading training in patients with juvenile maculopathy – Evaluation of two training programs. Vision 2002 Conference Abstract Book (p. 117).
19 Nilsson UL, Frennesson C, Nilsson SEG: Location and stability of a newly established eccentric retinal fixation locus suitable for reading, achieved through training of patients with a dense scotoma. Optom Vis Sci 1998;12:873–878.
20 Laubengaier C, Trauzettel-Klosinski S, Sadowski B Spectrum and effectivity of low vision care in the Low Vision Clinic Tübingen. Invest Ophthalmol Vis Sci 1997;38:841.

Orientation

21 Krastel H, Moreland JD: Cut-off filters and the acuity luminance function in retinitis pigmentosa; in Zrenner E, Krastel H, Goebel HH (eds): Advances in the Biosciences. Oxford, Pergamon, 1987, vol 68, pp 129–135.
22 Wolffsohn JS, Cochrane AL, Khoo H, Yoshimitsu Y, Wu S: Contrast is enhanced by yellow lenses because of selective reduction of short-wavelength light. Optom Vis Sci 2000;77:73–81.
23 Krastel H, Jaeger W, Blankenagel A: Complete and incomplete achromatopsia. Folia Ophthalmol 1989;14:269–274.
24 Eperiesi F, Fowler CW, Evans BJ: Do tinted lenses or filters improve visual performance in low vision? A review of the literature. Ophthalmic Physiol Opt 2002;22:68–77.
25 Pelli DG, Robson JG, Wilkins AJ: The design of a new letter chart for measuring contrast sensitivity. Clin Vision Sci 1988;2:187–199.
26 Rüther K, von Gruben C, Zrenner E: Wirkung von Kantenfiltern auf die Kontrastwahrnehmung bei Retinitis pigmentosa. Fortschr Ophthalmol 1991;88:829–832.
27 Wetzel C, Auffarth GU, Krastel H, Blankenagel A, Alexandridis E: Verbesserung der Kontrastempfindlichkeit bei hoher Adaptationsleuchtdichte durch Kantenfilter bei Retinitis pigmentosa. Ophthalmologe 1996;93:456–462.
28 Gallasch G, Krastel H, Hofmeister S: Early disorders of retinal function in diabetes mellitus. Ophthalmologe 1992;89:395–399.
29 Berson EL, Mehaffey L, Rabin AR: A night vision pocketscope for patients with retinitis pigmentosa. Design considerations. Arch Ophthalmol 1974;91:495–500.
30 Morrissette DL, Marmor MF, Goodrich GL: An evaluation of night-vision mobility aids. Ophthalmology 1983 90:1226–1230.
31 Friedburg C, Serey L, Sharpe L, Trauzettel-Klosinski S, Zrenner E: Evaluation of the night vision spectacles on probands with impaired vision. Graefes Arch Clin Exp Ophthalmol 1999; 237:125.
32 Rohrschneider K, Spandau U, Wechsle S, Blankenagel A: Einsatz einer neuen Nachtsichtbrille. Klin Monatsbl Augenheilkd 2000;217:88–93.
33 Rohrschneider K, Bruder I, Aust R, Blankenagel A: LVES – A new opto-electronic low vision aid: first results. Klin Monatsbl Augenheilkd 1997;210:105–110.
34 Weckerle P, Trauzettel-Klosinski S, Kamin G, Zrenner E: Task performance with the Low Vision Enhancement System (LVES). Vis Impair Res 2000;2:155–162.

35 Spaulding SJ, Patla AE, Elliott DB, Flanagan J, Rietdyk S, Brown S: Waterloo Vision and Mobility Study: Gait adaptations to altered surfaces in individuals with age-related maculopathy. Optom Vis Sci 1994;71:770–777.
36 Sunanto J, Nakata H: Indirect tactual discrimination of heights by blind and blindfolded sighted subjects. Percept Mot Skills 1998;86:383–386.
37 Bitjoka L, Pourcelot L: New blind mobility aid devices based on the ultrasonic Doppler effect. Int J Rehabil Res 1999;22:227–231.
38 Kooijman AC, Uyar MT: Walking speed of visually impaired people with two talking electronic travel systems. Vis Impair Res 2000;2:81–93.
39 Stanton D, Foreman N, Wilson PN: Uses of virtual reality in clinical training: Developing the spatial skills of children with mobility impairments. Stud Health Technol Inform 1998;58:219–232.

Prof. Dr. Susanne Trauzettel-Klosinski
Department of Pathophysiology of Vision and Neuro-Ophthalmology,
University Eye Hospital
Schleichstrasse 16, D–72076 Tübingen (Germany)
E-Mail setrauze@med.uni-tuebingen.de

Author Index

Subject Index